WIND ELECTRICAL SYSTEMS

S.N. Bhadra
*Professor, College of Engineering and
Management, Kolaghat, West Bengal*

D. Kastha
*Associate Professor, Department of Electrical Engineering,
Indian Institute of Technology Kharagpur, West Bengal*

S. Banerjee
*Professor, Department of Electrical Engineering,
Indian Institute of Technology Kharagpur, West Bengal*

OXFORD
UNIVERSITY PRESS

OXFORD
UNIVERSITY PRESS

YMCA Library Building, Jai Singh Road, New Delhi 110001

Oxford University Press is a department of the University of Oxford.
It furthers the University's objective of excellence in research, scholarship,
and education by publishing worldwide in

Oxford   New York
Auckland   Cape Town   Dar es Salaam   Hong Kong   Karachi
Kuala Lumpur   Madrid   Melbourne   Mexico City   Nairobi
New Delhi   Shanghai   Taipei   Toronto

With offices in
Argentina   Austria   Brazil   Chile   Czech Republic   France   Greece
Guatemala   Hungary   Italy   Japan   Poland   Portugal   Singapore
South Korea   Switzerland   Thailand   Turkey   Ukraine   Vietnam

Oxford is a registered trade mark of Oxford University Press
in the UK and in certain other countries.

Published in India
by Oxford University Press

© Oxford University Press 2005

The moral rights of the authors have been asserted.

Database right Oxford University Press (maker)

First published 2005

ISBN-13: 978-0-19-567093-6
ISBN-10: 0-19-567093-0

Typeset in Computer Modern
by Archetype, New Delhi 110063
Printed in India by Chaman Enterprises, Delhi 110002
and published by Manzar Khan, Oxford University Press
YMCA Library Building, Jai Singh Road, New Delhi 110001

To the memory of
Professor K.B. Menon and Professor K. Venkat Ratnam,
our teachers, former colleagues, and mentors,
who instilled among us a love for the teaching profession

Preface

Introduction

Over the past two decades, interest in the use of renewable energy resources has intensified. Of the various alternative energy sources, wind energy, being freely available and non-polluting, has been favoured the world over as the most feasible and economically competitive option.

A phased programme to develop wind energy in India started as early as 1985, and today the total installed capacity has reached 1650 MW, saving about 935,000 metric tons of coal. It goes without saying that such accelerated pace of development in this field needs, apart from R & D activities, trained manpower.

With wind energy harvesting becoming more and more important in the Indian energy scenario, it is essential for engineering students to be aware of the technologies involved in wind energy conversion. In keeping with the need of the hour, most engineering colleges in India have included a course on non-conventional power generation, either as a core subject or as an elective, in their curricula on electrical engineering, mechanical engineering, and energy engineering. Moreover, many young engineers, having graduated in mechanical or electrical engineering, are now finding employment in the emerging wind energy industry. They also need to prepare themselves for the challenges of the profession.

About the Book

The book aims at two segments of readership: BTech, BE, MTech, ME students as well as practising engineers and governmental bodies working on wind energy.

It first introduces the basics of wind energy conversion, and then concentrates on the issues of the conversion of wind energy into electrical energy, wind energy integration with the local grid, stand-alone generation and consumption, and hybrid power systems, where wind energy is integrated with other energy sources such as solar energy or diesel generators to provide reliable and continuous energy supply.

The subject of wind electric power conversion is multi-disciplinary in nature. It requires the knowledge of aerodynamics, wind turbines, electrical machines, power electronics, interfacing with solar/diesel power, etc. Teachers find it difficult to teach this subject in the absence of a comprehensive textbook dealing with wind power generation. In this book we have endeavoured to address this problem by putting together the fundamentals of all the necessary background subjects such as aerodynamics, electrical machines, and power electronics without assuming a background in electrical, mechanical, or aeronautical engineering.

Content and Structure

Chapter 1 starts with the issues related to the potential of wind as a source of power and the theoretical maximum efficiency attainable by wind turbines. It reviews the major types of wind turbines, concentrating on modern, low-solidity aerodynamic designs. It then introduces the aerodynamics of wind rotors and develops the expressions for torque, thrust, and aerodynamic power in terms of lift and drag forces. The desirable design criteria of propeller-type turbines and their control aspects are also discussed.

Chapter 2 deals with wind measurement and the statistical characteristics of the apparently random variation of wind speed. The output characteristics of wind machines combined with the statistical properties of wind speed distribution help to determine the potential of particular wind turbines at specific sites. The aspects of site and turbine selection are discussed.

Wind turbines convert wind energy into mechanical energy, which then needs to be converted into the electrical form using generators. In conventional generating stations, synchronous machines are used, while the variable-speed nature of wind energy necessitates a different strategy, wherein induction machines and synchronous machines (both wound rotor and permanent magnet types) are used in conjunction with power electronic converters. Chapter 3 presents, in a brief (but coherent) manner, the construction and performance features of various types of electrical generators—a knowledge of which is vital for appreciating the technical issues involved in the conversion of wind energy into electrical energy. Equivalent circuits, input–output relations, and d-q axis models are covered in this chapter.

Solid-state switching devices have brought about a great change in power engineering. The variable-speed operation of wind turbines has a lot of benefits, the most important being the possibility of maximizing the power output. However, the resulting variation of voltage and frequency with the variation of wind speed necessitates the use of power electronic converters in order to obtain good quality power output. Chapter 4 introduces power semiconductor devices and the various configurations of converters and inverters, including pulse width modulation and power-factor correction techniques. This chapter also shows how these techniques are implemented in various control topologies.

Wind power generation, as it stands today, is dominated by induction generators, of both the squirrel cage type and the wound rotor type. About 85% of the wind generators today are induction generators. Hence it was felt necessary to devote an entire chapter (Chapter 5) to the use of induction generators. The chapter describes the operation of constant-speed and variable-speed induction generators with grid-connected stator windings and self-excited induction generators. The chapter explains, through mathematical and circuit models, generating principles and presents the conditions under which these machines can generate and work effectively. Reactive power compensation and the effect of wind generators on a utility are briefly presented.

The present trend, particularly for high-capacity generators, is to adopt variable-speed operation of induction generators. Chapter 6 deals with the variable-speed operation of induction generators as well as wound rotor type and permanent magnet type synchronous generators, which are not directly connected to the grid. The chapter describes various control schemes and interfacing with the grid via power electronics from the viewpoint of variable-speed operation.

Wind being variable in nature, wind power alone cannot supply any utility continuously over a whole day and throughout a year. However, wind generators in conjunction with other sources such as solar and diesel generators, and storage devices such as batteries, can overcome this drawback. Chapter 7 discusses such 'hybrid' systems and presents their merits and limitations.

There have been rapid developments in the area of wind power generation, and much of the useful information is available in research papers and conference proceedings that are not easily accessible to teachers and students. Here, we have strived to incorporate recent developments (i.e., variable-speed, constant-frequency systems based on cage rotor and wound rotor induction machines, permanent magnet machines, hybrid systems, etc.) from all such sources in the form of a textbook. We have also included a good number of worked out and exercise problems on all topics.

Acknowledgements

This book would not have been possible without help from many quarters. First, we thank the Continuing Education Cell of the Indian Institute of Technology (IIT) Kharagpur for financially supporting this project. Next we thank Mr Sai of the Curriculum Development Cell, IIT Kharagpur, and Mr Somen Biswas of the Networking Laboratory, College of Engineering and Management, Kolaghat, for their help in preparing the figures. We are also grateful to Professor J.K. Das (Director) and Professor A.K. Chakraborty (Head) of the Electrical Engineering Department, College of Engineering and Management, Kolaghat, and Professor D.C. Saha, Head of the Electrical Engineering Department, IIT Kharagpur, for their encouragement. Finally, we wish to appreciate the patience shown by our families during the writing of the book. The content of this book has been taught to the students of energy engineering at IIT Kharagpur for the past few years, and their feedback has helped in improving it significantly. Last but not least, the authors do not hesitate to admit that the concepts and ideas contained in the papers listed in the Bibliography have been of immense help in writing this book. We would appreciate any feedback from the readers of this book towards the further improvement of its content and presentation.

SAILENDRA NATH BHADRA
DEBAPRASAD KASTHA
SOUMITRO BANERJEE

Contents

List of Symbols

In general, symbols have been defined in the text at first occurrence. However, this is a list of principal symbols for ready reference. Besides these, some symbols have been defined locally, either with word statements or with reference to figures. For time-varying quantities, lower-case letters refer to instantaneous values, whereas upper-case letters stand for average, rms, or maximum values.

a	axial interface factor
A	blade swept area, rotor area
C	dc link capacitor
C	capacitance
C_p	power coefficient (coefficient of performance)
C_T	torque coefficient
c	scale factor, length
D	duty ratio of a dc–dc converter
$d^e\text{-}q^e$	synchronously rotating direct- and quadrature-axis
$d^s\text{-}q^s$	stationary direct- and quadrature-axis
E_f	induced voltage phasor of a synchronous machine
E_{fm}	peak value of the excitation emf per phase of a synchronous machine
E_j	injected rotor emf of a slip-ring induction motor
E_r	air-gap voltage of a synchronous machine, residual voltage
F	ratio of actual frequency to base frequency, or per-unit frequency

F_D	drag force
F_L	lift force
f	frequency in hertz
f_b	base frequency
f'_f	field winding variables ($f \to v, i, \lambda$) referred to stator
f'_{dk}	direct-axis damper variables ($f \to v, i, \lambda$)
f'_{ds}	direct-axis stator variables ($f \to v, i$)
f'_{qk}	quadrature-axis damper variables ($f \to v, i, \lambda$)
f'_{qs}	quadrature-axis stator variables ($f \to v, i, \lambda$)
I	output load current, angle of inclination
I_a	armature current phasor of a synchronous machine
I_{dc}	dc output current of a bridge rectifier
I_d, i_d	dc output current of a phase-controlled rectifier
I_L	photogenerated current, latching current, load current
I_m	peak value of the balanced supply current
I_o	reverse saturation current of the photodiode, output current of a dc–dc converter
I_r	rotor current space-vector
I_s	stator current space-vector
i	angle of incidence (angle of attack)
i_{ds}	direct-axis stator current
i_{qs}	quadrature-axis stator current
i_{dr}, i_{qr}	direct-and quadrature-axis rotor currents
i'_{dr}	direct-axis rotor current (stator-referred)
i'_{qr}	quadrature-axis rotor current (stator-referred)
i_L	inductor current
J	moment of inertia of the motor load system
k	shape factor
k_{w_1}	stator phase winding factor of an induction motor
k_{w_2}	rotor phase winding factor of an induction motor
L''_d	direct-axis sub-transient reactance of a synchronous machine

L'_{dk}	direct-axis damper winding self-inductance referred to stator
L'_{qk}	quadrature-axis damper winding self-inductance referred to stator
L'_f	field winding self-inductance referred to stator
L_{md}	direct-axis magnetizing reactance
L_{mq}	quadrature-axis magnetizing reactance
m_a	modulation index of a voltage source inverter
m_d	fuel intake of a diesel engine
n_s	synchronous speed in rps
n_r	rotor speed in rps
P	generated power
P_{ag}	air-gap power of an induction motor
P_{Cu2}	rotor copper loss of an induction motor
P_m	effective mechanical power of an induction motor
P_S	maximum power output from a PV array
p	number of pole pairs
P_{max}	maximum extractable power
P_0	power contained in the wind
P_o	output power
R	speed regulation (when referred to diesel engine), rotor radius (when referred to wind turbine)
R_{dc}	dc link resistance
R'_f	field winding resistance referred to stator
R'_K	damper winding resistance referred to stator
R_r	per-phase rotor resistance of an induction motor
R'_r, X'_r	stator-referred rotor resistance and leakage reactance at the fundamental frequency
R_s	per-phase stator resistance (when referred to an induction motor), internal series resistance (when referred to a photovoltaic generator)
R_{sh}	shunt resistance of a photovoltaic generator
R_x	external rotor resistance of a slip-ring induction motor
s	per-unit slip of induction motor

T_{dm}	torque produced by a diesel engine
T_{el}	electrical load torque on a diesel engine
T_2	number of turns per phase of the rotor winding of an induction motor
T_s	switching time period of a dc–dc converter
T_{ON}	on period of a dc–dc converter
T_e	electromagnetic torque
T_1	number of turns per phase of the stator winding of an induction motor
T_1'	load torque
u	linear velocity of aerofoil
V_i	input dc voltage to a line-commutated inverter
$V_{L\text{-}L}$	line-to-line rms voltage of ac supply
V_t	terminal voltage phasor of a synchronous machine, carrier wave amplitude
V_m	peak value of the balanced supply voltage, sine wave amplitude
V_o	dc output voltage of a diode rectifier
V_{dc}	dc input voltage of a voltage source inverter
V_w	wind speed
V_∞	wind velocity without rotor interruption
v	wind velocity through the rotor
v_{abc}, i_{abc}	instantaneous phase voltages and currents
v_{ds}	direct-axis stator voltage
v_{qs}	quadrature-axis stator voltage
v'_{dr}	direct-axis rotor voltage (stator-referred)
v'_{qr}	quadrature-axis rotor voltage (stator-referred)
v_d	dc output voltage of a phase-controlled converter
$\bar{v}$	annual average wind speed
X_{ls}	per-phase stator leakage reactance of an induction motor
X_d	direct-axis synchronous reactance
X_q	quadrature-axis synchronous reactance
X_{lr}	per-phase rotor leakage reactance of an induction motor

X_m	magnetizing reactance of an induction motor
X_{mu}	unsaturated magnetizing reactance
X_s	stator self-reactance of an induction motor, synchronous reactance
X_r'	rotor self-reactance of an induction machine referred to the stator
X_σ	transient reactance of an induction motor
X_S	per-phase reactance of ac source
Y	admittance
α	firing angle of a phase controlled converter, pitch angle
β	advance angle of a phase-controlled converter
β	load angle of a voltage source inverter, torque angle
ϵ	combustion efficiency of a diesel engine, angle
η_a	aerodynamic efficiency
θ_e	instantaneous position of the d-axis of the synchronous machine
θ_{eo}	initial value of θ_e
ϕ_f	field flux of a synchronous machine
ϕ_{ar}	armature reaction flux of a synchronous machine
ϕ_{al}	leakage flux of a synchronous machine
λ	solar insolation
λ_0	tip speed ratio
λ_{ds}	direct-axis stator flux linkage
λ_{qs}	quadrature-axis stator flux linkage
λ_{dr}'	direct-axis rotor flux linkage (stator-referred)
λ_{qr}'	quadrature-axis rotor flux linkage (stator-referred)
λ_{ds}^s	d-axis component of the stator flux linkage in the stationary reference frame
λ_{qs}^s	q-axis component of the stator flux linkage in the stationary reference frame
μ	overlap angle of a line-commutated converter
ϕ	displacement angle in a phase-controlled converter
ϕ_m	maximum air-gap flux per pole

ρ	air density
σ	total leakage coefficient
τ_G	diesel engine speed governor time constant
τ_1	diesel time delay of a diesel engine
ω	angular frequency of ac supply, generator speed
ω_e	synchronous speed in rad/s
ω_r	rotor speed in rad/s
ω_{sl}	slip speed in elec. rad/s

1   Fundamentals of Wind Turbines

Wind energy is one of the most available and exploitable forms of renewable energy. Winds blow from a region of higher atmospheric pressure to one of lower atmospheric pressure. The difference in pressure is caused by (a) the fact that the earth's surface is not uniformly heated by the sun and (b) the earth's rotation. Essentially, wind energy is a by-product of solar energy, available in the form of that kinetic energy of air.

Wind has been known to man as a natural source of mechanical power for long. The technology of wind power has evolved over this long period. Of the various renewable energy sources, wind energy has emerged as the most viable source of electrical power, and is economically competitive with the conventional sources.

1.1   Historical Background

Before the invention of that steam engine, wind power had been used for centuries in sailing ships and then for pumping water and grinding grain. With the development of the steam engine and due to the economic advantages of the exploitation of fossil fuels, effort towards the development of the means of utilizing wind energy took a back seat in many countries. In the late nineteenth century, as electricity was used to transmit and consume energy, with thermal and hydel power becoming the favoured sources, wind energy fell further into disfavour.

However, some countries lacked adequate fuel and water-power resources, which led them to look for alternative ways of generating electricity. Denmark was such a country. It pioneered the development of windmills for the generation of electricity in the 1890s. Since then there has been an increasing endeavour to build wind-driven generators for use in isolated communities with negligible resources in the form of water power or coal, and at places which cannot be economically connected to public supply networks. Another reason for the worldwide interest in developing wind power plants is the rapidly increasing demand for electrical energy and the consequent depletion of fossil fuels, namely, oil and coal, whose reserves are limited. The depletion of reserves, increase in demand, and certain factors in world politics have together contributed to a sharp rise in the cost of thermal power generation. Many places also do not have the potential for generating hydel power. Nuclear power generation was once treated with great optimism, but with the knowledge of the environmental hazards associated with possible leakage from nuclear plants, most countries have decided not to install them anymore.

The growing awareness of these problems led to heightened research efforts for developing alternative sources of energy for generation of electricity. The most desirable source would be one that is non-pollutant, available in abundance, and renewable, and can be harnessed at an acceptable cost in both large-scale and small-scale systems. The most promising source satisfying all these requirements is wind, a natural energy source. Wind energy conversion may be mechanical or electrical in nature, but the present focus is on electricity generation. The maximum extractable energy from the 0–100 m layer of air has been estimated to be of the order of 10^{12} kWh per annum, which is of the same order as hydroelectric potential.

The development of wind energy for electrical power generation got a boost when, in the early decades of the twentieth century, aviation technology resulted in an improved understanding of the forces acting on blades moving through air. This resulted in the development of turbines with two or three blades. High speed and high efficiency of turbines were the necessary conditions for successful electricity generation. As the impending energy crisis became evident after the Second World War, governments in many European countries gave serious thought to wind energy as a viable supplementary source. Worldwide, interest in wind energy thus grew. Through the efforts of countless scientists and engineers from various disciplines, wind energy has now matured as a economically viable renewable source of energy.

1.2   Power Contained in Wind

The power contained in wind is given by the kinetic energy of the flowing air mass per unit time. That is,

$$P_0 = \frac{1}{2}(\text{air mass per unit time})(\text{wind velocity})^2$$

$$= \frac{1}{2}(\rho A V_\infty)(V_\infty)^2$$

$$= \frac{1}{2}\rho A V_\infty^3 \tag{1.1}$$

where P_0 is the power contained in wind (in watts), ρ is the air density ($\simeq 1.225$ kg/m^3 at 15 °C and normal pressure), A is the rotor area in (square metre), and V_∞ is the wind velocity without rotor interference, i.e., ideally at infinite distance from the rotor (in metres per second).

1.3   Thermodynamics of Wind Energy

It is easy to appreciate the fact that though kinetic in nature, wind is low-quality energy. It is basically a relatively unidirectional motion of air molecules, in that not all the molecules move in the same direction. There is random and disorderly thermal motion of the molecules in all directions. Only the algebraic summation yields a resultant value in one direction. Naturally, the order or organization of this form of energy is low in comparison with the motion of a shaft, where all the molecules share a common motion. Our objective in wind energy conversion is to transform this energy into the rotation of a shaft or the flow of electrons; only then it becomes useful to mankind.

The second law of thermodynamics states that whenever there is a transformation from low-quality energy to high-quality energy, it is impossible to achieve 100 per cent efficiency even in theory. There is always a theoretical maximum limit on the efficiency. In the case of conversion of heat energy into mechanical energy, the limit is given by $1 - T_2/T_1$, where T_1 and T_2 are the temperatures of the source and the sink, respectively, expressed in absolute scale. Similarly, in the case of conversion of wind energy into the mechanical energy of a rotating shaft, there must be some theoretical upper limit on the efficiency of the conversion. What is that upper limit? We discuss this limit in the following section.

1.4   Efficiency Limit for Wind Energy Conversion

In order to find the maximum efficiency ideally attainable in a wind machine, let us consider an ideal converter in the form of a disc of area A which extracts a fraction of the power contained in the wind flowing through it (Fig. 1.1). The velocity of the incoming air unaffected by rotor interference is V_∞, that of the air passing through the disc is v, and that at infinite distance away from the disc is V_2. The pressures of the incoming and outgoing air at infinite distance from the disc are the same at P_∞, but there is a pressure difference $(P^+ - P^-)$ between the two sides of the disc.

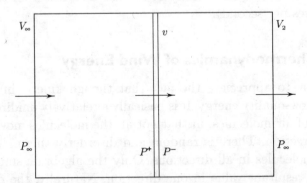

Fig. 1.1   The passage of wind through an ideal converter

It is assumed that the flow is axial, and that no rotational kinetic energy is imparted to the air stream. The flow through the disc separates distinctly from the remainder of the air stream.

Applying Bernoulli's theorem for the air streams on the two sides of the disc, we get, respectively,

$$\frac{1}{2}\rho V_\infty^2 + P_\infty = \frac{1}{2}\rho v^2 + P^+ \tag{1.2}$$

and

$$\frac{1}{2}\rho V_2^2 + P_\infty = \frac{1}{2}\rho v^2 + P^- \tag{1.3}$$

Subtracting, we get

$$P^+ - P^- = \frac{1}{2}\rho(V_\infty^2 - V_2^2) \tag{1.4}$$

The thrust on the disc is given by the area multiplied by the pressure difference, i.e.,

$$T = A(P^+ - P^-)$$

$$= \frac{1}{2}\rho A(V_\infty^2 - V_2^2) \tag{1.5}$$

The thrust is also given by

$$T = \dot{m}(V_\infty - V_2) = \rho A v(V_\infty - V_2) \tag{1.6}$$

Equating the two expressions for the thrust, we get

$$\frac{1}{2}\rho A(V_\infty^2 - V_2^2) = \rho A v(V_\infty - V_2)$$

or,

$$v = \frac{1}{2}(V_\infty + V_2) \tag{1.7}$$

It is conventional to work out the axial interference factor a in terms of

$$v = V_\infty(1 - a) \tag{1.8}$$

For $a = 0$ there is no interference, and so $v = V_\infty$. At $a = 1$ there is complete blockade of the flow of wind, and $v = 0$. For a normal wind turbine, a will take some value between 0 and 1. Substituting the expression for v, we are able to express V_2 also in terms of a as

$$V_\infty(1 - a) = \frac{1}{2}(V_\infty + V_2)$$

$$V_2 = V_\infty(1 - 2a) \tag{1.9}$$

Power extraction is given by the drop in kinetic energy of the air stream per unit time:

$$P_1 = \frac{1}{2}\rho Av(V_\infty^2 - V_2^2) \tag{1.10}$$

Substituting for v and V_2, we get

$$P_1 = 2\rho AV_\infty^3 a(1 - a)^2 \tag{1.11}$$

Thus the power output is a non-linear function of a. At the two extreme values $a = 0$ and 1, the power output is zero. Therefore the power output should reach a maximum for some value of a between 0 and 1. To find this value of a, we differentiate P_1 with respect to a and equate it to zero to get

$$\frac{dP_1}{da} = 2\rho AV_\infty^3(1 - 4a + 3a^2) = 0 \tag{1.12}$$

This quadratic equation has two solutions at $a = 1, 1/3$. Here $a = 1$ would mean that $v = 0$, which is impossible. So only $a = 1/3$ is physically acceptable. This gives the maximum extractable power as

$$P_{\text{max}} = \frac{8}{27}\rho AV_\infty^3 = \frac{16}{27}P_0 \tag{1.13}$$

which is reached when $v = 2V_\infty/3$.

This means that the theoretical maximum power extractable from wind is $16/27$ times the power contained in wind. This limit, first proved by Albert Betz in 1919, is called the *Betz limit*.

1.5   Maximum Energy Obtainable for a Thrust-operated Converter

There is a category of wind energy converters which operate on the thrust of wind on a solid surface. In such cases, the maximum power extractable is not given by the Betz limit.

1.5.1   Efficiency Limit for a Thrust-operated Converter

When wind strikes a surface blocking its way, it exerts a force on the surface. If the surface is stationary, the magnitude of this force, given by the change in the momentum of the wind, is $C_F(\rho AV_\infty^2)$, where C_F is the force coefficient and A is the area of the surface. However, in this case the power absorbed by the surface is zero. If the surface is moving with a velocity u, the force is given by $C_F[\rho A(V_\infty - u)^2]$, and the power absorbed by the surface by $C_F[\rho A(V_\infty - u)^2]u$.

It is apparent that the magnitude of the power extracted depends on the speed at which the surface is allowed to move and one can expect a maximum point to exist with respect to this speed. To find this maximum, we differentiate the above power function with respect to u and obtain

$$\frac{dP_1}{du} = \rho A C_F(V_\infty^2 - 4V_\infty u + 3u^2) \tag{1.14}$$

Equating to zero and taking $u/V_\infty = \beta$, we get

$$1 - 4\beta + 3\beta^2 = 0 \tag{1.15}$$

This equation has solutions at $\beta = 1, 1/3$, of which only $\beta = 1/3$ is feasible. Substituting this value in the expression for power extraction, we get a maximum:

$$P_{\max} = C_F \frac{8}{27}\left(\frac{1}{2}\rho A V_\infty^3\right) \tag{1.16}$$

In an ideal condition the force coefficient becomes unity, i.e., the wind transfers all its lost kinetic energy to the surface. Even in such a case, the maximum power extracted is only 8/27 times the power contained in the wind. This implies that a high-efficiency electricity-producing wind turbine should not operate on the principle of thrust force. In fact, all such turbines utilize aerodynamic forces, which allows one to get turbine efficiencies very close to the Betz limit of 16/27. However, thrust operated wind turbines are still in use where efficiency is not the prime requirement.

1.6   Types of Wind Energy Conversion Devices

Wind energy conversion devices can be broadly categorized into two types according to their axis alignment.

(1) *Horizontal-axis wind turbines* can be further divided into three types.
    (i) 'Dutch-type' grain-grinding windmills
    (ii) Multiblade water-pumping windmills

(iii) High-speed propeller type windmills
(2) *Vertical-axis wind turbines* come in two different designs.
 (i) The Savonius rotor
 (ii) The Darrieus rotor

These devices are described in detail in the following sections.

1.6.1 'Dutch' Windmills

Man has used Dutch windmills for a long time. In fact, the grain-grinding windmills that were widely used in Europe since the middle ages were Dutch. These windmills operated on the thrust exerted by wind. The blades, generally four, were inclined at an angle to the plane of rotation. The wind, being deflected by the blades, exerted a force in the direction of rotation. The blades were made of sails or wooden slats (Fig. 1.2).

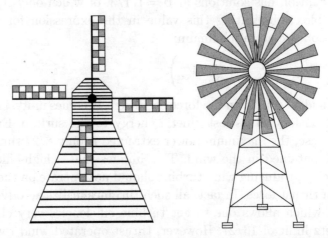

Fig. 1.2 The 'Dutch' windmill and the multiblade water-pumping windmill

In the early stages of the development of windmills, orienting the blades in the direction of the wind was accomplished manually, usually with the help of a tail pole. Later the 'fan-tail' system was introduced, in which there was a small windmill behind and at right angles to the main one, directly driving the orientation system. When the main windmill faced the wind, the fan-tail did not. When the wind direction changed, the fan-tail rotated and turned the main windmill back to the wind.

The Dutch windmill became obsolete with the advent of cheaper fossil fuels and because operating it requires high-skilled labour. However, with the depletion of fossil fuels there has been a renewed urge to exploit renewable energy resources. It was found that certain modifications in the earlier design of windmills make them very useful for pumping water.

1.6.2 Multiblade Water-pumping Windmills

Modern water-pumping windmills have a large number of blades— generally wooden or metallic slats—driving a reciprocating pump. As the mill has to be placed directly over the well, the criterion for site selection concerns water availability and not windiness. Therefore, the mill must be able to operate at slow winds. The large number of blades give a high torque, required for driving a centrifugal pump, even at low winds. Hence sometimes these are called *fan-mills* (see Fig. 1.2).

As these windmills are supposed to be installed at remote places, mostly as single units, reliability, sturdiness, and low cost are the prime criteria, not efficiency. The blades are made of flat steel plates, working on the thrust of wind. These are hinged to a metal ring to ensure structural strength, and the low speed of rotation adds to the reliability. The orientation is generally achieved by a *tail-vane*.

These machines should have an inbuilt protection against high winds and storms. This may be achieved simply by mounting the orienting tail-vane slightly off the axis of the main rotor. The windmill orientation then depends on a combination of the axial thrust of the wind on the rotor and the thrust on the tail-vane. The latter dominates at low winds, orienting the rotor almost in the direction of the wind. But the former dominates at high winds and makes the rotor face away from the wind.

1.6.3 High-speed Propeller-type Wind Machines

The horizontal-axis wind turbines that are used today for electrical power generation do not operate on thrust force. They depend mainly on the aerodynamic forces that develop when wind flows around a blade of aerofoil design. As has been shown already, windmills working on thrust force are inherently less efficient.

To understand how a modern electricity-producing wind turbine operates, let us first take a look at how an aerofoil works. Suppose an aerofoil—say an aeroplane wing—is moving in a stream of wind as shown in Fig. 1.3. The wind stream at the top of the aerofoil has to traverse a longer path than that at the bottom, leading to a difference in velocities. This gives rise to a difference in pressure (Bernoulli's principle), from which a *lift force* results. There is of course another force that tries to push the aerofoil back in the direction of the wind. This is called the *drag force*. The aggregate force on the aerofoil is then determined by the resultant of these two forces (Fig. 1.4).

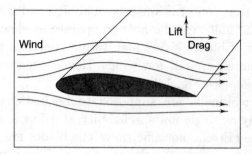

Fig. 1.3 Flow of wind over an aerofoil blade

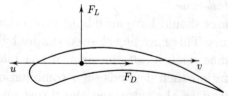

Fig. 1.4 The aerodynamic forces in an aerofoil moving in the direction of the wind; u is the aerofoil velocity, v is the wind velocity, F_L is the lift force, and F_D is the drag force

In this example, the aerofoil and the wind move along the same line. What happens when they do not? Then these forces are determined by the wind speed as seen by the aerofoil, called the *relative wind*. This is given by the vector summation of the wind velocity and the negative of the aerofoil velocity. The lift force F_L will now be perpendicular to the relative wind, and the drag force

F_D parallel to it. The magnitude of these two forces will also be proportional to that of the relative wind (see Fig. 1.5).

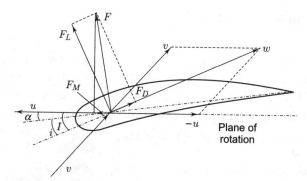

Fig. 1.5 The relative wind direction and the aerodynamic forces. Note the production of the resultant force F_M along the direction of motion. w is the relative wind direction, F_L is the lift force, F_D is the drag force, F_M is the moment force, I is the inclination angle, i is the incidence angle, and α is the pitch angle

We can see from Fig. 1.5 that the lift force and the drag force have opposing components along the direction of motion. If the lift force dominates the drag, there will be a resultant force along the direction of motion, giving a positive push to it. In fact, this is the force that creates the torque in a modern wind turbine. The blades are of aerofoil section, which move along the stream of wind. They are so aligned that the drag force is minimized and the lift force maximized, and this gives the blades a net positive torque (see Fig. 1.6).

Fig. 1.6 Arrangement and angle of blades in a propeller-type wind turbine

There will of course be another component of the two forces— that perpendicular to the direction of blade motion. This force is called the *thrust force*. This component tries to topple the tower and is a problem at high wind speeds.

Such horizontal-axis wind turbines are efficient and very suitable for electrical power generation.

1.6.4 The Savonius Rotor

The Savonius rotor is an extremely simple vertical-axis device that works entirely because of the thrust force of wind. The basic equipment is a drum cut into two halves vertically. The two parts are attached to the two opposite sides of a vertical shaft. As the wind blowing into the structure meets with two dissimilar surfaces—one convex and the other concave—the forces exerted on the two surfaces are different, which gives the rotor a torque. By providing a certain amount of overlap between the two drums, the torque can be increased. This is because the wind blowing into the concave surface turns around and gives a push to the inner surface of the other drum, partly cancelling the wind thrust on the convex side (see Fig. 1.7). It has been found that an overlap of about one-third the drum diameter gives the optimum result.

Fig. 1.7 The Savonius rotor

The Savonius rotor is inexpensive and simple, and the material required for it is generally available in any rural area, enabling on-site construction of such windmills. However, its utility is limited to pumping water because of its relatively low efficiency.

1.6.5 The Darrieus Rotor

In 1931, a vertical-axis device for wind energy conversion was invented by G.J. Darrieus of the United States, but was forgotten for a long time. The energy crisis renewed interest in windmill development in the 1970s, which reinvented the use of the Darrieus rotor for wind energy conversion.

The peculiarity of the Darrieus rotor is that its working is not at all evident from its appearance. Two or more flexible blades are attached to a vertical shaft as shown in Fig. 1.8. The blades bow outward, taking approximately the shape of a parabola, and are of symmetrical aerofoil section. At first sight it appears that the forces on the blades at the two sides of the shaft should be the same, producing no torque! Then how does it work?

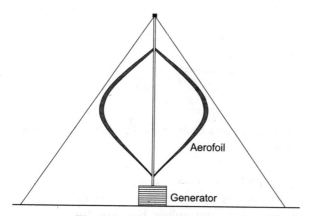

Fig. 1.8 The Darrieus rotor

In fact, the torque is zero when the rotor is stationary. It develops a positive torque only when it is already rotating. This means that such a rotor has no starting torque and has to be started using some external means.

The principle of operation is shown in Fig. 1.9. One blade of the rotor is shown in four successive positions along the path of rotation. At each position the blade velocity vector u, the wind velocity vector v, the relative wind w, the lift force F_L, and the drag force F_D are shown. It can be seen that at each position the lift force has a positive component in the direction of rotation, giving rise to a net positive torque.

This torque is not the same in all the positions. It varies from zero when the blade is moving directly upwind or downwind to a maximum about a quarter of a revolution later. The torque thus

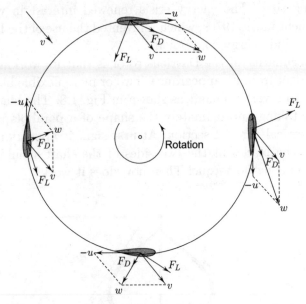

Fig. 1.9 Forces acting on a blade of the Darrieus rotor at four successive positions

makes two complete excursions from zero to maximum and back in each revolution—both in the positive sense. The pulsations in the shaft torque can be minimized by using three blades. However, the two-blade design has the advantage of a lower erection cost because it can be assembled flat on the ground and erected as a unit.

The torque is also a function of the speed of rotation and the wind speed. We have mentioned earlier that at zero rotational speed, the torque is also zero. It increases with the rotational speed. Naturally, Darrieus rotors are designed for a high speed of rotation. The torque increases with wind speed up to a certain value and then falls off at very high wind speeds. This means that this design has an inbuilt protection from stormy weather—the rotor tends to stall at high winds.

The curvature of the blade, a 'troposkein' in geometrical parlance, is simply the shape assumed by a flexible weightless chord spinning with its two ends tied at two points. It can be closely approximated by a parabola:

$$\left(\frac{z}{a}\right)^2 = 1 - \frac{r}{b} \tag{1.17}$$

Figure 1.10 explains the parameters appearing in this equation. The swept area is given by $A = 2DH/3$, where $D = 2b$ and $H = 2a$ are the diameter and the height of the rotor, respectively.

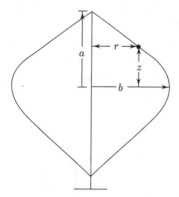

Fig. 1.10 The shape of a Darrieus rotor blade

As the curvature of the blade corresponds to that of a completely flexible chord, the only stress that can develop in the blade is tension. Blade failure can be avoided simply by ensuring a high tensile strength along the direction of the blade. Fibre-reinforced materials with fibres aligned along the blade are quite suitable for its construction.

As the Darrieus rotor operates on lift force, its efficiency approaches that of modern horizontal-axis propeller-type windmills. The theoretical limit of power extraction, under certain assumptions, can be shown to be 0.554 times the power contained in wind; the corresponding Betz limit for a horizontal-axis machine is 0.593.

The Darrieus rotor, with its high efficiency and high speed, is perfectly suited for electrical power generation. The cost of construction is low because the generator and the gear assembly can be located at ground level, drastically reducing the cost of the tower. However, then it is unable to take advantage of the high wind speeds available at higher altitudes.

The starting torque is generally provided by an electrical machine, which initially runs as a motor but later changes to the generator mode as the Darrieus rotor starts generating power. Occasionally, two Savonius rotors are fitted at the two ends of the vertical shaft to provide the starting torque.

A variant of this machine, using the same concept, is the Giromill shown in Fig. 1.11, in which the blades are straight, resulting in simple construction. However, in such a case the centrifugal force developed in the blade will produce stress, trying to bend it. The blades have to be strong enough in the transverse direction to withstand this stress. Moreover, the vertical shaft cannot be secured with guywires, and so the coupling at the base has to be strong enough to keep it vertical when subjected to strong winds.

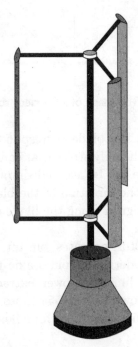

Fig. 1.11   The straight-blade vertical-axis turbine—the Giromill

1.7   Some Relevant Definitions

Before proceeding further, one should get acquainted with the terms frequently used in the literature on wind energy.

Solidity

The solidity of a wind rotor is the ratio of the projected blade area to the area of the wind intercepted. The projected blade area does not mean the actual blade area; it is the blade area met by the wind or projected in the direction of the wind.

The solidity of the Savonius rotor is naturally unity, as the wind sees no free passage through it. For a multiblade water-pumping windmill, it is typically around 0.7. For high-speed horizontal-axis machines, it lies between 0.01 and 0.1; for the Darrieus rotor also it is of the same order.

Solidity has a direct relationship with torque and speed. High-solidity rotors have high torque and low speed, and are suitable for pumping water. Low-solidity rotors, on the other hand, have high speed and low torque, and are typically suited for electrical power generation.

Tip Speed Ratio

The tip speed ratio (TSR) of a wind turbine is defined as

$$\lambda = \frac{2\pi RN}{V_\infty} \tag{1.18}$$

where λ is the TSR (non-dimensional), R is the radius of the swept area (in metres), N is the rotational speed in revolutions per second, and V_∞ is the wind speed without rotor interruption (in metres per second).

The TSRs of the Savonius rotor and the multiblade water-pumping windmills are generally low. In high-speed horizontal-axis rotors and Darrieus rotors, the outer tip actually turns much faster than the wind speed owing to the aerodynamic shape. Consequently, the TSR can be as high as 9. It can be said that high-solidity rotors have, in general, low TSRs and vice versa.

Power Coefficient

The power coefficient of a wind energy converter is given by

$$C_p = \frac{\text{power output from the wind machine}}{\text{power contained in wind}} \tag{1.19}$$

The power coefficient differs from the efficiency of a wind machine in the sense that the latter includes the losses in mechanical transmission, electrical generation, etc., whereas the former is just the efficiency of conversion of wind energy into mechanical energy of the shaft. In high-speed horizontal-axis machines, the theoretical maximum power coefficient is given by the Betz limit.

The graph of the power coefficient against the TSR is a very important yardstick in the characterization of wind energy converters. Typical curves for different types of windmills are shown in Fig. 1.12.

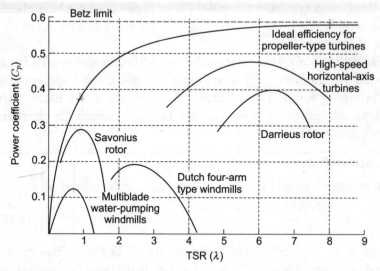

Fig. 1.12   Curves of C_p versus TSR for different types of windmills

Wind Turbine Ratings and Specifications

Since the same wind turbine can produce widely varying amounts of electrical power depending on the wind speed, a standard procedure must be evolved to specify the rating of a machine. Some manufacturers specify the power rating along with the wind speed at which the power rating is obtained. However, since there is no clear-cut rule to define what the rated wind speed should be, it works to the advantage of the manufacturer to specify a speed on the higher side, showing a higher power rating.

A more meaningful specification is the combination of the rotor diameter and the peak power rating of the generator. One index used to compare various wind turbine designs is the *specific rated capacity* (SRC), defined as

$$\text{SRC} = \frac{\text{power rating of the generator}}{\text{rotor swept area}}$$

The SRC varies between 0.2 for small rotors to 0.6 for large ones.

1.8 Aerodynamics of Wind Rotors

In this section we will study the aerodynamics of an aerofoil and develop equations for the various forces acting on a windmill blade. This exposition is aimed at enabling the reader to appreciate the logical structure with which various forces and stresses can be estimated and preliminary designs made. The actual design procedures used in the industry involve many complications. In the present text we avoid burdening the reader with such details and concentrate only on the essential concepts.

From now on we will mainly concentrate on the high-speed propeller-type wind turbine with electrical power generation in view. Before analysing the aerodynamic forces, one should be conversant with the relevant definitions and conventions. See Figs 1.5 and 1.13 for the diagrammatic representations of the velocities and forces.

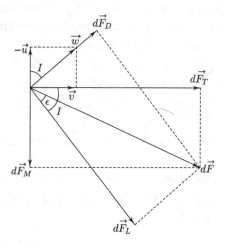

Fig. 1.13 Vector diagram of the velocities and forces, the blade vector $\vec{R}$ is normal to the plane of this diagram

Axial speed of wind   Speed of wind through the rotor in metres per second, denoted by $\vec{v}$

Speed of blade element   The speed of a blade element at a distance r from the rotor axis is $2\pi r N$ in metres per second, denoted by $\vec{u}$

Relative velocity   The velocity of air flow relative to the blade, $\vec{w} = \vec{v} - \vec{u}$

Blade axis The longitudinal axis going through the blade; it is possible to vary the inclination of the blade relative to the plane of rotation around this axis

Blade section at r The intersection of the blade with a cylinder of radius r, whose axis is the rotor axis; the section is aerofoil-shaped

Pitch angle The angle α between the chord of the aerofoil section at r and the plane of rotation, also called the *setting angle*

Angle of inclination The angle between the relative velocity vector and the plane of rotation, denoted by I

Angle of incidence The angle of incidence is the angle between the relative velocity vector and the chord line of the aerofoil, denoted by i. It is clear that $i = I - \alpha$. It is also called the *angle of attack.*

Lift force The lift force is the component of the aerodynamic force in the direction perpendicular to the relative wind. It is given by $F_L = \rho A_b w^2 C_L/2$ newtons, where C_L is the dimensionless *lift coefficient* and A_b is the blade area in square metres.

Drag force The component of the aerodynamic force in the direction of the relative wind; it is given by $F_D = \rho A_b w^2 C_D/2$ newtons, where C_D is the *drag coefficient*

Total aerodynamic force The total aerodynamic force on a blade element is given by $\vec{F} = \vec{F_L} + \vec{F_D}$

Thrust force The component of $\vec{F}$ along the direction of wind, denoted by $\vec{F_T}$

Torque force The component of $\vec{F}$ along $\vec{u}$, denoted by $\vec{F_M}$

Aerodynamic moment The moment of $\vec{F}$ about the axis in newton metres, denoted by M

1.8.1 Analysis Using the Blade-element Theory

Consider a blade element of length dr at a distance r from the rotor axis as shown in Fig. 1.14. The magnitudes of the lift and drag forces developed in this blade element are given by

$$dF_L = \frac{1}{2}\rho \, dA_b \, w^2 C_L \tag{1.20}$$

and

$$dF_D = \frac{1}{2}\rho \, dA_b \, w^2 C_D \tag{1.21}$$

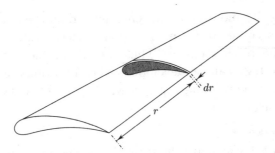

Fig. 1.14 A blade element

Their resultant gives the total aerodynamic force dF, which can be resolved into the axial thrust dF_T and the moment-producing force dF_M. The aerodynamic moment is given by r multiplied by dF_M. From Fig. 1.13, we see that

$$dF_T = dF_L \cos I + dF_D \sin I \quad \text{∠ between velocity \& plane} \tag{1.22}$$

$$dF_M = dF_L \sin I - dF_D \cos I \tag{1.23}$$

$$dM = r(dF_L \sin I - dF_D \cos I) \tag{1.24}$$

If the angular velocity is ω, we can write

$$w^2 = u^2 + v^2 = r^2\omega^2 + v^2 \tag{1.25}$$

where $\quad u = \omega r = v \cot I \tag{1.26}$

Substituting Eqns (1.20), (1.21), (1.25), and (1.26) in Eqns (1.22) and (1.24), we get

$$dF_T = \frac{1}{2}\rho \, dA_b \, v^2 (1 + \cot^2 I)(C_L \cos I + C_D \sin I) \tag{1.27}$$

and

$$dM = \frac{1}{2}\rho \, dA_b \, v^2 r (1 + \cot^2 I)(C_L \sin I - C_D \cos I) \tag{1.28}$$

The power developed (in watts) is given by

$$dP = \omega dM = \frac{1}{2}\rho \, dA_b \, v^3 \cot I \operatorname{cosec} I (C_L \sin I - C_D \cos I)$$

$$\tag{1.29}$$

It is clear that the torque and aerodynamic power depend on the angle I, which, in turn, depends on the wind speed and rotational speed. The lift and drag coefficients depend on the angle of attack i, which is equal to $I - \alpha$. These coefficients can be varied by varying the pitch angle α. Thus, based on these equations, a control strategy can be developed, whereby a maximum amount of power can be produced at any wind speed by suitably varying the pitch angle.

Example 1.1 Calculate the total thrust and aerodynamic power developed in a three-blade wind turbine at a wind velocity (V_∞) of 9 m/s. The machine specifications are as follows.

> Diameter = 9 m
> Rotational speed = 100 rpm
> Blade length = 4 m
> TSR = 5.23
> Chord length = 0.45 m, uniform throughout the blade
> Pitch angle $\alpha = 5°$, no twist
> Distance from axis to inner edge of blade = 0.5 m
> Aerofoil section = NACA 23018

Solution To apply the blade element theory, we have to divide the length of the blade into smaller segments. Breaking it up into very small segments would give better results, but the computation would be cumbersome. It is a good exercise for students to write computer programs to do the calculations.

However, to demonstrate the method, we divide the blade into four equal sections of length 1 m each. The values related to the middle of each section are denoted by the corresponding subscript. The area of each section is $0.45 \times 1.0 = 0.45$ m^2.

From the Betz theory we know that maximum efficiency is obtained when $v = 2V_\infty/3$. Assuming this condition, we take $v = 6$ m/s. The rotational speed is $100/60 = 1.66$ rps. As $\tan I = v/u$, the values of I at each section are

$$I_1 = \tan^{-1} \frac{6}{2 \times \pi \times 1.0 \times 1.66} = 29.81°$$

$$I_2 = \tan^{-1} \frac{6}{2 \times \pi \times 2.0 \times 1.66} = 15.98°$$

$$I_3 = \tan^{-1} \frac{6}{2 \times \pi \times 3.0 \times 1.66} = 10.81°$$

$$I_4 = \tan^{-1} \frac{6}{2 \times \pi \times 4.0 \times 1.66} = 8.15°$$

As $i = I - \alpha$, we get $i_1 = 24.81°$, $i_2 = 10.98°$, $i_3 = 5.81°$, and $i_4 = 3.15°$.

From the characteristic curves of the NACA 23018 aerofoil, we get

$$C_{L1} = 0.95, \quad C_{D1} = 0.0105$$
$$C_{L2} = 1.20, \quad C_{D2} = 0.0143$$

$$C_{L3} = 0.75, \quad C_{D3} - 0.0092$$
$$C_{L4} = 0.46, \quad C_{D4} = 0.0078$$

Equation (1.29) gives

$$dP_1 = 0.5 \times 1.25 \times 0.45 \times 6^3 \times \cot 29.81(1 + \cot^2 29.81)$$
$$\times (0.95 \sin 29.81 - 0.0105 \cos 29.81)$$
$$= 198.72 \text{ W}$$

Similarly, $dP_2 = 886.14$ W, $dP_3 = 1190.52$ W, and $dP_4 = 1213.38$ W. So the total power $= 3 \times (198.72 + 886.14 + 1190.52 + 1213.38) = 10,466$ W ≈ 10 kW.

The thrusts are calculated as follows:

$$dF_{T1} = 0.5 \times 1.25 \times 0.45 \times 6^2 \times (1 + \cot^2 I)$$
$$\times (0.95 \cos I + 0.0105 \sin I)$$
$$= 33.98 \text{ N}$$

Similarly, $dF_{T2} = 154.24$ N, $dF_{T3} = 212.94$ N, and $dF_{T4} = 229.96$ N. Therefore, the total thrust that the tower has to withstand at rated wind speed is $3 \times (33.98 + 154.24 + 212.94 + 229.96) = 1893.4$ N.

1.8.2 Aerodynamic Efficiency

The aerodynamic efficiency of a blade element is defined as

$$\eta_a = \frac{\text{useful power extracted from the wind}}{\text{power supplied by the wind}}$$

$$= \frac{\vec{u} \cdot d\vec{F}}{\vec{v} \cdot d\vec{F}}$$

$$= \frac{u\, dF_M}{v\, dF_T}$$

$$= \frac{u(dF_L \sin I - dF_D \cos I)}{v(dF_L \cos I + dF_D \sin I)}$$

Dividing the numerator and the denominator by $dF_L \sin I$, we get

$$\eta_a = \frac{u}{v} \frac{[1 - (dF_D/dF_L)\cot I]}{(\cot I + dF_D/dF_L)} \tag{1.30}$$

Let ϵ be the angle between $d\vec{F}_L$ and $d\vec{F}$. Then,

$$\tan \epsilon = \frac{dF_D}{dF_L} = \frac{C_D}{C_L} \tag{1.31}$$

By simplifying Eqn (1.30) using Eqns (1.26) and (1.31), we get

$$\eta_a = \cot I \, \frac{1 - \tan \epsilon \cot I}{\cot I + \tan \epsilon}$$

$$= \frac{1 - \tan \epsilon \cot I}{1 + \tan \epsilon \tan I} \tag{1.32}$$

Equation (1.32) provides the clue for designing a blade. It shows that the lower the value of $\tan \epsilon$, the higher the efficiency. At the limit $\tan \epsilon$ equal to zero, the aerodynamic efficiency is unity. Of course this is not possible as it would mean zero drag force. However, we can conclude that it is desirable to have as low a value of ϵ as possible. But how small? This is generally determined from the plot of lift coefficient versus drag coefficient for the aerofoil section being used. These coefficients can be determined experimentally in a wind tunnel. Normally these data are available in the form of plots of C_L versus i and C_D versus i. The C_L versus C_D plot can be obtained from such data. A typical curve for various incidence angles, called the *Eiffel polar*, is given in Fig. 1.15 (see also the graphs at the end of this chapter).

It can be seen that the angle ϵ is minimum for a particular incidence angle i, where the ϵ-line [shown as a dashed line in Fig. 1.15(a)] becomes tangential to the Eiffel polar. For this angle of incidence, the aerodynamic efficiency is maximum. Thus this curve gives the optimal values of both ϵ and i for any particular aerofoil section.

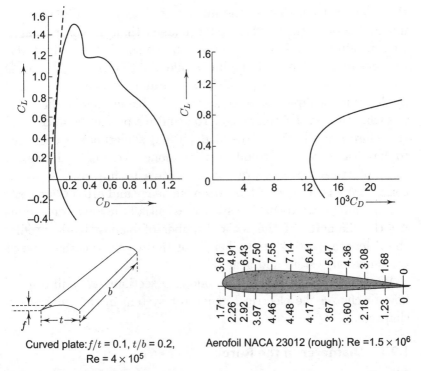

Curved plate: $f/t = 0.1$, $t/b = 0.2$, Re $= 4 \times 10^5$

Aerofoil NACA 23012 (rough): Re $=1.5 \times 10^6$

Fig. 1.15 Eiffel polar plots for a curved plate and an aerofoil (Wortmann 1978)

1.9 Design of the Wind Turbine Rotor

From the preceding section it should be clear that efficient design of a blade maximizes the lift and minimizes the drag. Minimization of the drag means that the aerofoil should face the relative wind in such a way that the smallest possible area is exposed to the wind drag. The angle of the relative wind is determined by the relative magnitudes of u and v. The wind velocity is constant throughout the rotor area, but the blade velocity increases from the inner edge to the tip. So, neither the magnitude nor the angle of the relative wind is constant throughout the length of the blade.

From this observation, a few qualitative conclusions about the desirable features of a wind turbine blade can be drawn. First, as the aerofoil has to face the relative wind at all points, it should have a varying pitch angle along the blade. In other words, it should have a twist. Second, as the lift force developed at the tip is higher than that at the inner edge due to the difference in magnitude of w, the tip tries to move faster than the central parts. This produces a stress that may cause blade failure. Moreover, the tension developed at the inner side due to centrifugal force is more than that at the tip. Both these problems can be remedied by designing a tapering blade so that the blade area at the tip is less than that at the inner edge. To achieve a perfect balance of all these forces at all parts of the blade, it may sometimes be desirable to have a varying aerofoil section along the blade. However, some of these features are too expensive to implement, and a compromise is often struck between efficiency and production cost.

Designing a wind turbine system essentially involves determining the diameter of the rotor, number of blades, blade profile, chord length, setting angle, height of the tower, and the type of transmission system and gear box.

We will study all this in the following sections. We will discuss how to choose the electrical generator system in the subsequent chapters.

1.9.1   Diameter of the Rotor

The diameter of the rotor is determined from the operating wind speed and the rated power output. The generated power is given by

$$P = P_0 \eta_e \eta_m C_p$$
$$= \frac{1}{2} \rho A V_\infty^3 \eta_e \eta_m C_p$$
$$= \frac{1}{8} \pi \rho D^2 V_\infty^3 \eta_e \eta_m C_p \tag{1.33}$$

where η_m is the efficiency of mechanical transmission and η_e is the efficiency of electrical generation. If the rated P (W), V_∞ (m/s), and C_p are known, the diameter in metres can be found out.

In the absence of the above data, the following simple formulae can be used for the initial estimation of the maximum aerodynamic power:

$$P = 0.15 D^2 V_\infty^3 \quad \text{for slow rotors} \tag{1.34}$$
$$P = 0.20 D^2 V_\infty^3 \quad \text{for fast rotors} \tag{1.35}$$

Example 1.2 Find the required diameter of a wind turbine to generate 4 kW at a wind speed of 7 m/s and a rotor speed of 120 rpm. Assume power coefficient = 0.4, efficiency of mechanical transmission = 0.9, and efficiency of generator = 0.95.

Solution   Equation (1.33) gives

$$4.0 = \frac{1}{8} \times 0.95 \times 0.9 \times 0.4 \times \pi \times 1.25 \times 7^3 D^2$$

or

$$D = 8.33 \text{ m}$$

1.9.2   Choice of the Number of Blades

It is obvious that the efficiency of power extraction from wind depends on the proper choice of the number of blades. There will be little power extraction if the blades are so close to each other or rotate so fast that every blade moves into the turbulent air created by the preceding blade. It will also be less than optimum if the blades are so far apart or move so slowly that much of the air stream passes through the wind turbine without interacting with a blade. Thus, the choice of the number of blades should depend on the TSR.

Let t_a be the time taken by one blade to move into the position previously occupied by the preceding blade. For an n-bladed rotor rotating at an angular velocity ω,

$$t_a = \frac{2\pi}{n\omega} \tag{1.36}$$

Let t_b be the time taken by the disturbed wind, caused by the interference of a blade, to move away and normal air to be re-established. It depends on the wind speed v and the length of the strongly perturbed air stream, say d. This length, stretching both upwind and downwind, depends on the size and shape of the blades.

$$t_b = \frac{d}{v} \tag{1.37}$$

For maximum power extraction, t_a and t_b should be more or less equal. From Eqns (1.36) and (1.37), we have

$$\frac{\omega}{v} \approx \frac{2\pi}{nd}$$

The choice of the number of blades therefore depends on the value of d, which has to be determined empirically. A large number of blades implies high solidity—hence high torque and low speed. On the other hand, a small number of blades implies low torque and high speed. Therefore, a large number of blades are used in wind turbines meant for pumping water or other mechanical functions that require a high starting torque. For modern electricity-generating wind turbines, the empirical measurement of d and the requirement of a high TSR lead to a small number of blades, generally only two or three.

Though both two-blade and three-blade designs are equally popular, their choice depends on certain factors. The two-blade designs have less nacelle weight and are much simpler to erect. Three-blade turbines involve 33% more weight and cost, though the power coefficient increases only by 5–10 %. On the other hand, the three-blade design has smoother power output and a more balanced gyroscopic force, and therefore less blade fatigue and less chances of failure.

1.9.3   Choice of the Blade Profile and Material

For low-TSR water-pumping windmills, the blade is generally a flat metallic plate. In some cases it is a simple, circularly curved metallic sheet, which leads to certain aerofoil-like characteristics, but with uniform thickness throughout the blade. Because of their low rigidity, these blades have to be fixed to a circular metallic frame for structural support. At low rotational speeds, the expenditure for fabricating aerofoil blades does not justify itself.

However, for high-speed wind turbines, the blade profile must have an aerofoil section. A systematic study of the characteristics of various aerofoil sections has been done by the National Advisory Committee for Aeronautics (NACA) of USA and the University of Göttingen in Germany for use in the construction of aeroplane wings and propellers. The geometries of the various profiles and their aerodynamic characteristics are available from the NACA series and the Göttingen series of publications (for example,

NACA 4412, 4415, 4418, 23012, 23015, 23018 and Göttingen 623, 624, etc.). As the aerodynamic characteristics of wind turbine blades are essentially the same as those for aeroplane wings and propellers, a choice can be made from the profiles given in these publications.

In choosing the appropriate profile, caution must be exercised. The studies presented in these literatures are generally for high Reynolds number flows[1] and are not valid for lower Reynolds numbers. The Reynolds number for the flow of wind around a wind turbine blade is generally much lower than that around an aeroplane wing. Hence the aerofoil performance of a wind turbine may deviate significantly from that expected upon the application of the characteristic values obtained from these studies. A special series for aerofoils particularly suitable for low Reynolds number flows has been published by Wortmann from the University of Stuttgart, Germany, and by the Applied Aerodynamics Group, University of Illinois at Urbana-Champaign (available at http://amber.aae.uiuc.edu/~m-selig). These are more useful for wind turbine design.

High-speed turbine blades are made of high-density wood or fibreglass and epoxy composites. Among all the components of a wind turbine, the blades have the maximum likelihood of failure because of mechanical stresses. It is therefore important to keep the stresses within limits. This is achieved by limiting the rotor speed to a predesigned limit, stalling the turbine at high wind speeds, and restricting the acceleration and deceleration rates.

1.9.4 Determination of the Blade Chord

The width of the blade chord can be determined by a simple manipulation of the relationships developed in the previous sections. For a blade element between radii r and $r + dr$, we can state that

---

[1]The Reynolds number of any fluid flow is given by

$$\text{Re} = \frac{w b_p}{\nu'}$$

where b_p is the width of the profile (m), w is the flow velocity (m/s), and ν' is the kinematic viscosity (m^2/s). If the flow velocity is much higher than the viscosity, we get a high Reynolds number—the condition prevailing in an aeroplane propeller. On the other hand, if the flow around the blade is slower, the viscosity term becomes significant and we get a low Reynolds number flow.

(see Fig. 1.13)

$$dF = \frac{dF_L}{\cos \epsilon}$$

and

$$w = \frac{v}{\sin I}$$

Now we can obtain the thrust produced by the blade element as

$$dF_T = dF \cos(I - \epsilon)$$

$$= dF_L \frac{\cos(I - \epsilon)}{\cos \epsilon}$$

$$= \frac{1}{2} \rho C_L dA_b w^2 \frac{\cos(I - \epsilon)}{\cos \epsilon}$$

$$= \frac{1}{2} \rho C_L dA_b \frac{v^2 \cos(I - \epsilon)}{\sin^2 I \cos \epsilon}$$

If the blade chord length is c, the elemental blade area is given by $dA_b = c\,dr$. Thus the thrust on the blade element is given by

$$dF_T = \frac{1}{2} \rho C_L \frac{v^2 \cos(I - \epsilon)}{\sin^2 I \cos \epsilon} c\,dr \qquad (1.38)$$

Similarly, the expression for the moment produced by the blade element in terms of C_L, I, and ϵ becomes

$$dM = r\,dF_M$$

$$= r\,dF \sin(I - \epsilon)$$

$$= r\,dF_L \frac{\sin(I - \epsilon)}{\cos \epsilon}$$

$$= r \frac{1}{2} \rho C_L \frac{v^2 \sin(I - \epsilon)}{\sin^2 I \cos \epsilon} c\,dr \qquad (1.39)$$

The following factors should be borne in mind while determining the length of the blade chord.

(1) The total torque developed by the blade at the rated wind speed should be equal to the desired value.

(2) The moment and thrust produced by blade elements of equal length along the blade should be almost equal.

While designing a blade using the blade-element theory, the length of the blade is divided into a number of smaller elements within which the parameters are assumed to be constant. Parameters such as C_L, C_D, ϵ, and i are determined from aerofoil data sheets and the Eiffel polar plot of the aerofoil section. (Note that as $\tan \epsilon = C_D/C_L$, the tangent to the Eiffel polar gives the minimum ϵ only if C_D and C_L are plotted on the same scale.) I is given by the magnitudes of v and u at a particular blade element. The design parameters are computed for each element, keeping in mind that the summation of the forces over the whole blade length should yield the desired values. It is natural that the smaller the length of that blade element, the better the design. Computer programs can be developed easily to do this job.

It should be noted, however, that the complex variation of chord length computed above may not be the most desirable economically. In most cases a linear trapezoidal tapering is preferred for the sake of cost-effective manufacture.

1.9.5 Choice of the Pitch Angle

The pitch angle is given by $\alpha = I - i$. As I varies along the length of the blade, α should also vary to ensure an optimal angle of incidence at all points of the blade. Thus the desirable twist along the blade can be calculated easily.

The pitch angle should be such that $\tan \epsilon$ or C_D/C_L is minimum at all points of the rotor. Some authors recommend the use of the Eiffel polar plot for this purpose. In this method, the tangent to the Eiffel polar plot gives the minimum C_D/C_L. However, for the tangent to represent $\tan \epsilon$, both C_D and C_L should be drawn to the same scale. This becomes very inconvenient, since for most aerofoils, C_L is about two orders of magnitude higher than C_D. It is more convenient to plot the curve for C_D/C_L versus i. Its minimum point will then represent the optimal value of the incidence angle.

This method yields a twisted blade, that is, one that has different pitch angles at different distances from the axis. If the constraints in the production method do not permit a twist, the optimal value of α can be chosen for a suitable point on the blade, say $r = 0.8R$, and the same pitch angle maintained throughout the blade.

1.9.6 The Tower

In a horizontal-axis wind turbine, the tower supports the whole machinery, including the blades, the gear box, the generator, and the control equipment. It therefore requires high strength, which is achieved with a steel or concrete structure, based on tubular or lattice construction. It is necessary to avoid amplification of vibration through careful design of the resonant frequencies of the tower, blades, rotor, etc. vis-à-vis the wind fluctuation frequencies.

In general, for medium and large turbines, the height of the tower is slightly greater than the rotor diameter. Small turbines should have taller towers in comparison with their rotor diameters; otherwise the turbine would be too close to the ground surface and would experience poor wind speeds. Turbines with rated output between 10 kW and 100 kW have tower heights in the range of 20–30 m; 300-kW to 500-kW machines have towers 35 m to 40 m high.

1.9.7 The Transmission System and Gear Box

In general, the optimal speed of rotation of an electrical generator is much higher than the optimal speed of a wind turbine. In order to ensure that a low speed of the turbine produces a high rotational speed at the generator, a gear box is inserted in the transmission system. The arrangement inside the generator housing is schematically shown in Fig. 1.16.

If the gear system has a fixed gear ratio, the transmission system is relatively simple and inexpensive. However, in this case the efficiency suffers at low or high wind speeds. It has been found that for a particular site (with particular wind speed distribution characteristics), one particular choice of the gear ratio gives the highest system efficiency, and the curve falls off on both sides of this optimal gear ratio. Therefore, a judicious choice of the gear ratio is very important. Generally a speed ratio of 20–30 is chosen for wind electrical systems.

For a variable-speed wind turbine, a better overall efficiency may be obtained with a two-speed gear box which can switch from a low gear ratio at high wind speeds to a high gear ratio at low wind speeds so that the speed variation at the generator side is kept low.

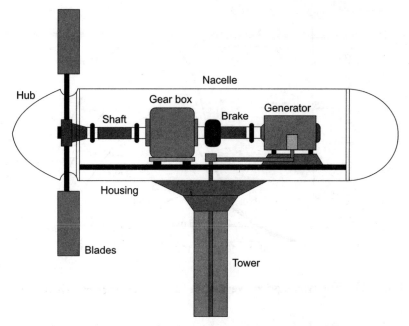

Fig. 1.16 The shaft–gear box–generator arrangement

1.10 Power–Speed Characteristics

The wind turbine power curves shown in Fig. 1.17 illustrate how the mechanical power that can be extracted from the wind depends on the rotor speed. For each wind speed there is an optimum turbine speed at which the extracted wind power at the shaft reaches its maximum. Such a family of wind turbine power curves can be represented by a single dimensionless characteristic curve, namely, the C_p-λ curve, as shown in Fig. 1.12, where the power coefficient is plotted against the TSR. For a given turbine, the power coefficient depends not only on the TSR but also on the blade pitch angle. Figure 1.18 shows the typical variation of the power coefficient with respect to the TSR λ with blade pitch control.

From Eqns (1.1) and (1.19), the mechanical power transmitted to the shaft is

$$P_m = \frac{1}{2}\rho C_p A V_\infty^3 \qquad (1.40)$$

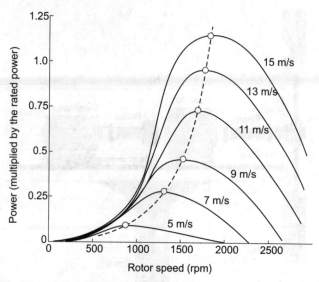

Fig. 1.17 The typical power versus speed characteristics of a wind turbine

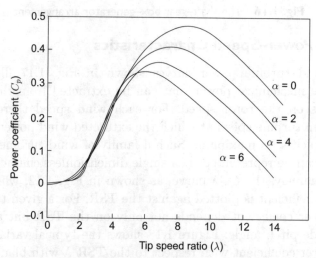

Fig. 1.18 Typical curves of power coefficient versus tip speed ratio for various values of the pitch angle α

where C_p is a function of the TSR λ and the pitch angle α. For a wind turbine with radius R, Eqn (1.40) can be expressed as

$$P_m = \frac{1}{2}\rho C_p \pi R^2 V_\infty^3 \tag{1.41}$$

For a given wind speed, the power extracted from the wind is maximized if C_p is maximized. The optimum value of C_p, say $C_{p,\text{opt}}$, always occurs at a definite value of λ, say λ_{opt}. This means that for varying wind speed, the rotor speed should be adjusted proportionally to adhere always to this value of $\lambda(=\lambda_{\text{opt}})$ for maximum mechanical power output from the turbine. Using the relation $\lambda = \omega R/V_\infty$ in Eqn (1.41), the maximum value of the shaft mechanical power for any wind speed can be expressed as

$$P_{\text{max}} = \frac{1}{2} C_{p,\text{opt}} \pi \left(\frac{R^5}{\lambda_{\text{opt}}^3} \right) \omega^3 \tag{1.42}$$

Thus the maximum mechanical power that can be extracted from wind is proportional to the cube of the rotor speed, i.e., $P_{\text{max}} \propto \omega^3$. This is shown in Fig. 1.17.

1.11   Torque–Speed Characteristics

Studying the torque versus rotational speed characteristics of any prime mover is very important for properly matching the load and ensuring stable operation of the electrical generator. The typical torque–speed characteristics of the two-blade propeller-type wind turbine, the Darrieus rotor, and the Savonius rotor are shown in Fig. 1.19. The profiles of the torque–speed curves shown in Fig. 1.19 follow from the power curves, since torque and power are related as follows:

$$T_m = \frac{P_m}{\omega} \tag{1.43}$$

From Eqn (1.42), at the optimum operating point $(C_{p,\text{opt}}, \lambda_{\text{opt}})$, the relation between aerodynamic torque and rotational speed is

$$T_m = \frac{1}{2} \rho C_{p,\text{opt}} \pi \left(\frac{R^5}{\lambda_{\text{opt}}^3} \right) \omega^2 \tag{1.44}$$

It is seen that at the optimum operating point on the C_p-λ curve, the torque is quadratically related to the rotational speed.

The curves in Fig. 1.19 show that for the propeller turbine and the Darrieus rotor, for any wind speed, the torque reaches a maximum value at a specific rotational speed, and this maximum shaft torque varies approximately as the square of the rotational speed. In the case of electricity production, the load torque depends on the electrical loading, and by properly choosing the load (or power electronic interface), the torque can be made to vary as the square of the rotational speed.

The choice of the constant of proportionality of the load is very important (see Fig. 1.20). At the optimal value, the load curve follows the maximum shaft power. But at a higher value, the load torque may exceed the turbine torque for most speeds.

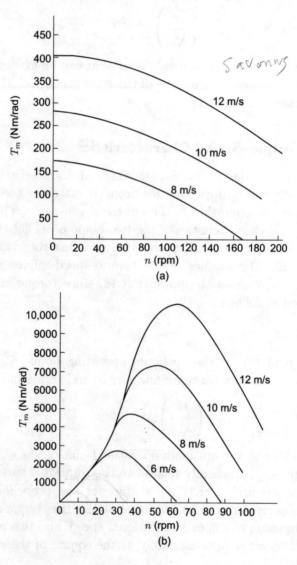

(a)

(b)

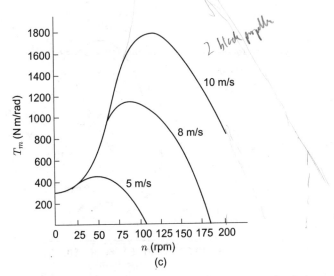

Fig. 1.19 The torque–speed characteristics of various types of wind machines: (a) the Savonius rotor, (b) the Darrieus rotor, (c) the two-blade propeller-type rotor

Consequently, the machine would fail to speed up above a very low value. If the constant K is lower than the optimum value, the machine may overspeed at the rated wind speed, activating the speed-limiting mechanism. Thus the proportionality constant of the load needs to be selected from a rather narrow range, about 10–20 % of the optimum power curve. Note that the point of maximum torque is not the same as that of maximum power.

As the power output is a product of torque and speed, it also has maxima that vary as the cube of the rotational speed. The matching characteristics of the load can make the load curve pass through the maximum power points at all wind speeds. For generators that feed power to the grid, the torque–speed characteristics are tuned using power electronic controls.

In terms of the power coefficient $C_p(\lambda, \alpha)$, the aerodynamic torque becomes

$$T_m = \frac{1}{2}\rho C_T \pi R^3 V_\infty^3 \tag{1.45}$$

where $C_T = C_p/\lambda$ is called the *torque coefficient*.

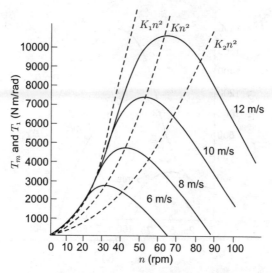

Fig. 1.20 The torque–speed characteristics of a wind turbine with $T \propto n^2$ load for different values of the proportionality constant K

1.12 Wind Turbine Control Systems

Wind turbines require certain control systems. Horizontal-axis wind turbines have to be oriented to face the wind. In high winds, it is desirable to reduce the drive train loads and protect the generator and the power electronic equipment from overloading, by limiting the turbine power to the rated value up to the furling speed. At gust speeds, the machine has to be stalled. At low and moderate wind speeds, the aim should be to capture power as efficiently as possible.

Along with many operating characteristics, the technical data sheet of a turbine mentions its output at a particular wind speed, generally known as the rated wind speed. This is the minimum wind speed at which the wind turbine produces its designated output power. For most turbines, this speed is normally between 9 and 16 m/s. The choice of the rated wind speed depends on the factors related to the wind speed characteristics of a given site, which are discussed in the next chapter. The generator rating is chosen so as to best utilize the mechanical output of the turbine at the rated wind speed.

Wind turbines can have four different types of control mechanisms, as discussed in the following.

1.12.1 Pitch Angle Control

This system changes the pitch angle of t.. blades according to the variation of wind speed. As discussed earlier, with pitch control, it is possible to achieve a high efficiency by continuously aligning the blade in the direction of the relative wind.

On a pitch-controlled machine, as the wind speed exceeds its rated speed, the blades are gradually turned about the longitudinal axis and out of the wind to increase the pitch angle. This reduces the aerodynamic efficiency of the rotor, and the rotor output power decreases. When the wind speed exceeds the safe limit for the system, the pitch angle is so changed that the power output reduces to zero and the machine shifts to the 'stall' mode. After the gust passes, the pitch angle is reset to the normal position and the turbine is restarted. At normal wind speeds, the blade pitch angle should ideally settle to a value at which the output power equals the rated power.

The pitch angle control principle is explained in Fig. 1.21. The input variable to the pitch controller is the error signal arising from the difference between the output electrical power and the reference power. The pitch controller operates the blade actuator to alter the pitch angle. During operation below the rated speed, the control system endeavours to pitch the blade at an angle that maximizes the rotor efficiency. The generator must be able to absorb the mechanical power output and deliver to the load. Hence, the generator output power needs to be simultaneously adjusted.

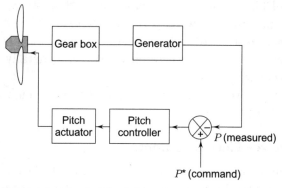

Fig. 1.21 The feedback loop for pitch angle control

Continuous pitch control is relatively expensive to incorporate, and the cost–benefit trade-off does not justify its use in small wind machines. However, the stalling mechanism must be incorporated to prevent damage of the turbine during turbulent weather conditions.

1.12.2 Stall Control

Passive stall control

Generally, stall control to limit the power output at high winds is applied to constant-pitch turbines driving induction generators connected to the network. The rotor speed is fixed by the network, allowing only 1–4 % variation. As the wind speed increases, the angle of attack also increases for a blade running at a near constant speed. Beyond a particular angle of attack, the lift force decreases, causing the rotor efficiency to drop. This is an intrinsic property and dispenses with the need for a complex control system and moving parts. The lift force can be further reduced to restrict the power output at high winds by properly shaping the rotor blade profile to create turbulence on the rotor blade side not facing the wind.

Active stall control

In this method of control, at high wind speeds, the blade is rotated by a few degrees in the direction opposite to that in a pitch-controlled machine. This increases the angle of attack, which can be controlled to keep the output power at its rated value at all high wind speeds below the furling speed. A passive controlled machine shows a drop in power at high winds. The action of active stall control is sometimes called *deep stall*. Owing to economic reasons, active pitch control is generally used only with high-capacity machines.

Figure 1.22 presents typical profiles of power curve for pitch control and stall control.

1.12.3 Power Electronic Control

In a system incorporating a power electronic interface between the generator and the load (or the grid), the electrical power delivered by the generator to the load can be dynamically controlled. The

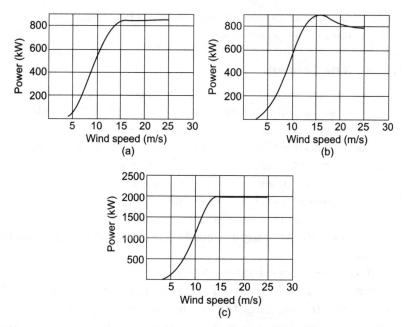

Fig. 1.22 Typical power profiles: (a) pitch control, (b) passive stall control, (c) active stall control

instantaneous difference between mechanical power and electrical power changes the rotor speed following the equation

$$J\frac{d\omega}{dt} = \frac{P_m - P_e}{\omega}$$ (1.46)

where J is the polar moment of inertia of the rotor, ω is the angular speed of the rotor, P_m is the mechanical power produced by the turbine, and P_e is the electrical power delivered to the load. Integrating Eqn (1.46), we get

$$\frac{1}{2}J\left(\omega_2^2 - \omega_1^2\right) = \int_{t_1}^{t_2} (P_m - P_e)dt$$

The advantage of this method of speed control is that it does not involve any mechanical action and is smooth in operation. A disadvantage is that fast variation of speed requires a large difference between the input power and output power, which scales as the moment of inertia of the rotor. This results in a large torque and hence increased stress on the blades. Moreover, continuous control of the rotor speed by this method implies continuous fluctuation of the power output to the grid, which is usually undesirable for the power system.

1.12.4   Yaw Control

This control orients the turbine continuously along the direction of wind flow. In small turbines this is achieved with a tail-vane. In large machines this can be achieved using motorized control systems activated either by a fan-tail (a small turbine mounted perpendicular to the main turbine) or, in case of wind farms, by a centralized instrument for the detection of the wind direction. It is also possible to achieve yaw control without any additional mechanism, simply by mounting the turbine downwind so that the thrust force automatically pushes the turbine in the direction of the wind.

The yaw control mechanism can also be used for speed control— the rotor is made to face away from the wind direction at high wind speeds, thereby reducing the mechanical power. However, this method is seldom used where pitch control is available, because of the stresses it produces on the rotor blades. Yawing often produces loud noise, and it is desirable to restrict the yawing rate in large machines to reduce the noise.

1.13   Control Strategy

For every wind turbine, there are five different ranges of wind speed, which require different speed control strategies (Fig. 1.23).

(a)   Below a cut-in speed, the machine does not produce power. If the rotor has a sufficient starting torque, it may start rotating below this wind speed. However, no power is extracted and the rotor rotates freely. In many modern designs the aerodynamic torque produced at the standstill condition is quite low and the rotor has to be started (by working the generator in the motor mode) at the cut-in wind speed.

(b)   At normal wind speeds, maximum power is extracted from wind. We have seen earlier that the maximum power point is achieved at a specific (constant) value of the TSR. Therefore, to track the maximum power point, the rotational speed has to be changed continuously in proportion to the wind speed.

(c)　At high winds, the rotor speed is limited to a maximum value depending on the design limit of the mechanical components. In this region, the C_p is lower than the maximum, and the power output is not proportional to the cube of the wind speed.

(d)　At even higher wind speeds, the power output is kept constant at the maximum value allowed by the electrical components.

(e)　At a certain cut-out or furling wind speed, the power generation is shut down and the rotation stopped in order to protect the system components.

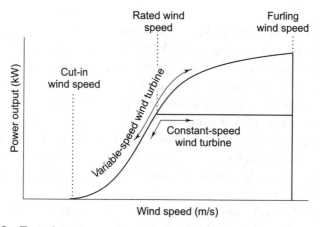

Fig. 1.23　Typical power versus wind speed characteristics of variable-speed wind machines

The last three control regimes can be realized with yaw control, pitch angle control (if these are installed), and eddy-current or mechanical brakes.

In the intermediate-speed range, the control strategy depends on the type of electrical power generating system used, and can be divided into two basic categories:

(a)　the constant-speed generation scheme, and

(b)　the variable-speed generation scheme.

The constant-speed generation scheme is necessary if the electrical system involves a grid-connected synchronous generator (the details are given in subsequent chapters). In the case of grid-connected squirrel cage induction generators, the allowable range of speed variation is very small, requiring an almost constant rotational speed.

However, constant-speed generation systems cannot maximize the extraction of the power contained in wind. We can see from Fig. 1.12 that the power coefficient reaches a maximum at a specific value of TSR for every type of wind turbine. Therefore, to extract the maximum amount of power from the wind, the turbine should operate at a constant TSR, which means that the rotational speed should be proportional to the wind speed. Hence the extraction of maximum power requires a variable-speed generation system with the speed control aimed at keeping a constant TSR.

Such systems can yield 20–30 % more power than constant-speed generation systems. With the development of induction generators and power electronic converters, designers are favouring variable-speed generation systems. We will, therefore, discuss the control strategies for such systems in greater detail.

The constant-TSR region, which encompasses the largest range of wind speeds, is generally achieved by regulating the mechanical power input through pitch control or the electrical power output by power electronic control. In many cases a combination of both is employed.

The control scheme generally takes two possible forms. In the first case, the value of the TSR for maximum C_p is stored in a microprocessor. The operating TSR is obtained from the measured values of the wind speed and rotational speed. An error signal is generated whenever the operating TSR deviates from the optimum TSR. If the current value of the TSR is greater than the optimum TSR, the power electronic converter increases the power output so that the rotational speed is reduced to the desired value. The opposite action is performed if the optimal value exceeds the current TSR.

This scheme has a few disadvantages. First, the wind speed measured in the neighbourhood of a wind turbine (or a wind farm) is not a reliable indicator of V_∞ because of the shadowing effects. Second, it is difficult to determine the value of TSR for maximum C_p. Third, this value changes during the lifetime of a wind turbine due to the changes in the smoothness of the blade surface, necessitating alterations in the reference setting.

A second control scheme is devised to continuously track the maximum power point (MPP) using the property that the C_p versus TSR curve has a single smooth maximum point. This means that if we operate at the maximum power point, small fluctuations in the rotational speed do not significantly change the power output, i.e., the MPP is characterized by $dP/d\omega = 0$. To implement this scheme, the speed is varied in small steps, the power output is measured, and $\Delta P/\Delta\omega$ is evaluated. If this ratio is positive, more mechanical power can be obtained by increasing the speed. Hence the electrical power output is decreased temporarily by power electronic control so that the speed increases. This increases the mechanical power, and the electrical power output is again raised to a higher value. The process continues until the optimum speed is reached, when mechanical power becomes insensitive to speed fluctuations. When the wind speed changes, this mechanism readjusts the speed at the optimum value.

While controlling the rotational speed, it should be remembered that a large difference between mechanical power and electrical power results in a large torque and, hence, a large stress on the rotor components (especially on the joints between the blades and the shaft). To avoid fatigue and failure, it is necessary to limit the acceleration and deceleration rates to values dictated by the structural strength of the mechanical parts.

The use of brakes

In the event of load tripping or accidental disconnection of the electrical load, the rotor speed may increase dangerously. This may even lead to the mechanical destruction of the rotor. Moreover, at very high wind speeds, the electrical power throughput has to be kept within limits to protect the generator and the power electronic converter. This can be done by reducing the rotational speed. However, this speed control cannot be achieved by power electronic control as discussed above, because that would call for an increase in the electrical power output—exactly the opposite of what was desired.

In these situations it is advisable to use brakes. Either an eddy-current or a mechanical brake (or a combination of these) is installed in most wind turbines. A mechanical brake is also necessary for stalling these turbines in gusty winds.

Summary

This chapter introduces the various types of wind energy conversion devices. It shows that thrust-operated wind turbines are inherently less efficient than lift-operated ones. The chapter then analyses in detail the working of the horizontal-axis propeller-type wind turbine using the blade-element theory. This theory is simple to understand and provides reasonable estimates of a turbine's characteristics. However, the reader should keep in mind that in actual practice the design of a turbine follows more complicated procedures, which are outside the scope of this introductory text.

The important mechanical characteristics of a wind turbine, such as power–speed characteristics and torque–speed characteristics, which will be needed for the later chapters, are also treated here. Finally, the control systems necessary for running a wind turbine are discussed.

Problems

1. A horizontal-axis wind turbine has a diameter of 5 m. When the wind speed unaffected by the turbine is 10 m/s, the turbine rotates at 300 rpm and produces 5 kW of mechanical power. Find the tip speed ratio and the power coefficient.

2. The turbine of Problem 1 is connected to an electrical generator and a variable load. The load is smoothly varied to obtain a maximum power output for the above wind condition. Find the wind speed through the turbine under that condition.

3. A Darrieus rotor has the following dimensions: $a = 2.5$ m, $b = 2$ m (see the illustration in Fig. 1.10). If it produces 3 kW of mechanical shaft power at $V_\infty = 10$ m/s, calculate the power coefficient.

4. A horizontal-axis wind turbine rotates at 120 rpm and the wind speed through the blade is 5.6 m/s. For a pitch angle of 8° (uniform throughout the blade), plot a graph showing the

variation of the angle of incidence i with the radial distance along a blade.

5. A horizontal-axis wind turbine has a blade length of 6 m and a total diameter of 14 m. The rated wind speed (V_∞) is 8 m/s and the rated tip speed ratio is 6.0. The chord length of 0.8 m and the pitch angle of 4° are uniform throughout the blade. Assuming the Reynolds number to be around 60,000, choose an aerofoil (from among those whose wind-tunnel test data are given in Figs 1.24–1.35)[2] that will give the maximum power output under the rated condition.

6. For the aerofoil BW-3 (at Reynolds number 59,400), draw the Eiffel polar plot and find the optimal value of the angle ϵ.

7. An electricity producing wind turbine has to rotate at 140 rpm for a rated wind speed of 8 m/s. What is the required diameter in order to produce 100 kW of electrical power? Assume $C_p = 0.43$, efficiency of mechanical transmission = 0.9, efficiency of electrical generator and power conditioning equipment = 0.92.

8. A two-blade wind turbine is to produce 300 kW of mechanical power. The blades are 9 m in length. Assuming three blade elements of equal length, find the optimal chord length for each blade element.

9. Consider a three-blade wind turbine that rotates at a constant speed of 100 rpm. The aerofoil is SA7035. The blade length is 6 m, the chord length is 0.8 m (uniform throughout the blade), the pitch angle α is 6° (no twist), and the distance from the axis to the inner edge of the blade is 1 m. Divide the blade into four elements and obtain the torque for various values of wind speed. Beyond which wind speed does the turbine stall?

10. Suppose the wind turbine of Problem 9 is made of aerofoil A18 and the pitch is controllable. At what angle should the blade be turned in order to produce a stall at $V_\infty = 18$ m/s?

---

[2] All the aerofoil data are taken from http://amber.aae.uiuc.edu/~m-selig

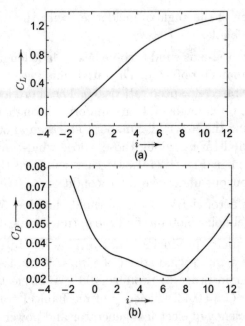

Fig. 1.24   The graphs of (a) lift coefficient and (b) drag coefficient versus angle of attack for the aerofoil BW-3 at Reynolds number 59,400

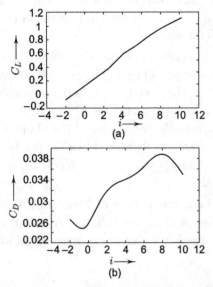

Fig. 1.25   The graphs of (a) lift coefficient and (b) drag coefficient versus angle of attack for the aerofoil Clark-Y at Reynolds number 60,700

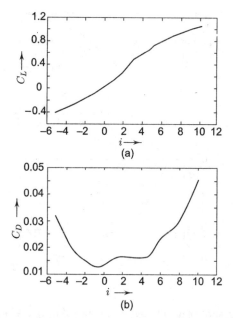

Fig. 1.26 The graphs of (a) lift coefficient and (b) drag coefficient versus angle of attack for the aerofoil RG14 at Reynolds number 60,400

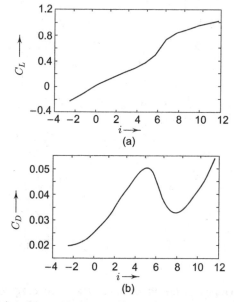

Fig. 1.27 The graphs of (a) lift coefficient and (b) drag coefficient versus angle of attack for the aerofoil E374 at Reynolds number 60,200

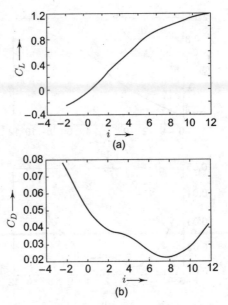

Fig. 1.28 The graphs of (a) lift coefficient and (b) drag coefficient versus angle of attack for the aerofoil Goe 417a at Reynolds number 59,300

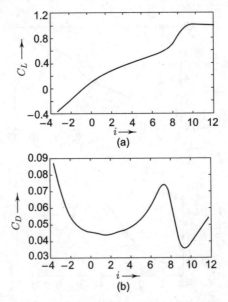

Fig. 1.29 The graphs of (a) lift coefficient and (b) drag coefficient versus angle of attack for the aerofoil LRN1007 (B) at Reynolds number 59,500

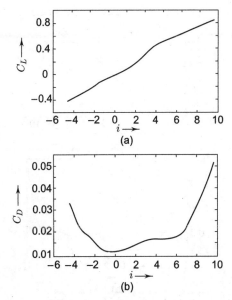

Fig. 1.30 The graphs of (a) lift coefficient and (b) drag coefficient versus angle of attack for the aerofoil S6063 at Reynolds number 60,000

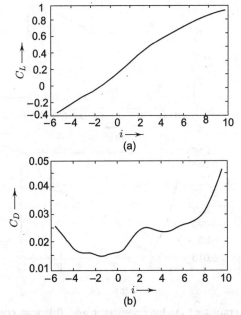

Fig. 1.31 The graphs of (a) lift coefficient and (b) drag coefficient versus angle of attack for the aerofoil S7012 at Reynolds number 59,938.

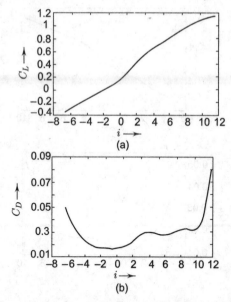

Fig. 1.32 The graphs of (a) lift coefficient and (b) drag coefficient versus angle of attack for the aerofoil SA7035 at Reynolds number 59,893

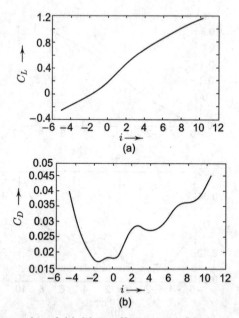

Fig. 1.33 The graphs of (a) lift coefficient and (b) drag coefficient versus angle of attack for the aerofoil SD7037 (D) at Reynolds number 59,200

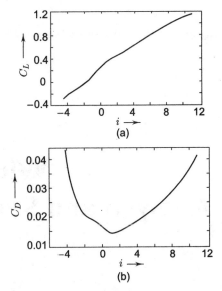

Fig. 1.34 The graphs of (a) lift coefficient and (b) drag coefficient versus angle of attack for the aerofoil A18 at Reynolds number 103,018

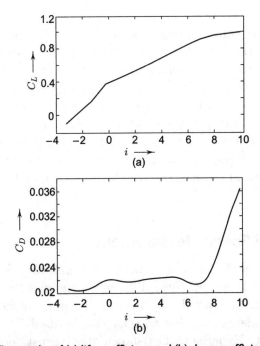

Fig. 1.35 The graphs of (a) lift coefficient and (b) drag coefficient versus angle of attack for the aerofoil Avistar at Reynolds number 100,000

2

Wind Site Analysis and Selection

It follows from common sense that a site suitable for the installation of wind turbines should be 'windy'. However, the windiness of a site needs to be specified in quantitative terms. In order to do so, one needs to first obtain data on wind speeds and directions by installing measuring instruments at potential sites. Since wind speed and direction vary continuously, estimation of the power generation potential requires some statistical analysis. Finally, one has to obtain a proper match between the characteristics of the wind turbine and those of the site. We will take up these issues in this chapter.

2.1 Wind Speed Measurements

The device used for wind speed measurement is called an *anemometer*. There are three different techniques for wind speed measurement. In general, any measurable phenomenon that has strong dependence on wind velocity can be used for wind speed measurement. Experience has shown that thrust, pressure, and the cooling effect, are the three most convenient parameters using which wind speed can be directly measured.

2.1.1 Robinson Cup Anemometer

The Robinson cup anemometer consists of a vertical shaft carrying three or four horizontal arms, at the ends of which there are hemispherical cups of thin sheet metal. The circular rims of the cups are in vertical planes passing through the common axis of rotation. The thrust of wind is greater on the concave sides than on the convex ones, thereby leading to the rotation of the vertical shaft (Fig. 2.1).

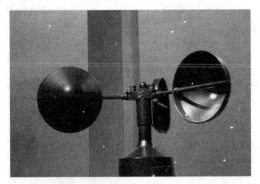

Fig. 2.1 The Robinson cup anemometer

As this is a vertical-axis device, there is no problem of orientation along the wind direction. The wind velocity has a linear relationship with the speed of rotation, which is measured by a photocell-operated digital counter. The display can be precalibrated to give the wind speed directly. Modern devices have facilities for continuous data logging and storage, from which data can be retrieved later for analysis.

At very low wind speeds, the readings of the cup anemometer can be erroneous due to the friction of the bearings. During fast variations of wind speed, the inertia effect may be significant; e.g., when the wind speed drops quickly, the anemometer tends to rotate faster and takes time to slow down. In spite of these minor drawbacks, the Robinson cup anemometer is the most extensively used instrument for wind speed measurement.

2.1.2 Pressure Tube Anemometer

The pressure tube anemometer a simple mechanical device suitable for stand-alone application in remote windy locations.

Structure-wise, it has two distinct parts. The head, which is usually mounted on a mast at the desired height, consists of a horizontal tube, bent at one end and supported by two concentric vertical tubes. The horizontal end is connected to the inner tube. At the other side of the outer tube, there are a few holes a little below the horizontal tube. The entire head is free to rotate, which is turned to face the wind by a vane.

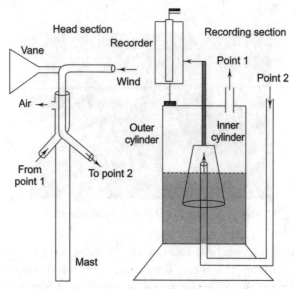

Fig. 2.2 The pressure tube anemometer

The wind blowing into the horizontal tube creates pressure, which is communicated through a flexible tube to a recorder. Again, the wind blowing over the small holes in the concentric tube creates a suction effect, which is also communicated to the recorder through a second flexible tube.

In the recording apparatus, a copper vessel, closed at one end, floats inverted in a cylindrical metal container partly filled with water and sealed from the outside air. The wind pressure from the horizontal tube of the head is transmitted to the space inside the float, causing it to rise as wind blows. This is assisted by the suction that is applied to the space above the float.

Thus, as the wind speed rises and falls, the float also rises and falls, and its motion is transferred to a pen tracing a record on a sheet of paper by means of a rod passing through an airtight passage at the top of the cylinder. If the paper movement is spring operated, the device does not need any electrical supply.

The float can be so shaped, in accordance with the law relating pressure to wind velocity, that the velocity scale on the chart is linear. Most pressure tube anemometers also have wind direction recorders taking signals from the tail-vane, so that both the speed and direction of wind are recorded.

2.1.3 Hot Wire Anemometer

A hot wire anemometer uses the cooling effect of wind on an electrically heated platinum or tungsten wire to measure wind velocities. The wire is heated by a constant-current source. With the variation of wind speed, the wire temperature varies, which varies the resistance of the wire. Naturally, in order to find the wind speed, it suffices to measure the resistance of the wire using any standard method. The calibration has to take into account the resistance–temperature characteristics of the wire and the ambient temperature of air.

In a hot wire anemometer, the temperature difference between the wire and the ambient air is inversely proportional to the square root of the wind velocity:

$$T_w - T_a \propto \frac{1}{\sqrt{v}} \tag{2.1}$$

Because of this relation, this anemometer is useful especially for measuring small wind velocities.

2.2 Wind Speed Statistics

Since the power contained in wind varies with the cube of the wind speed, the average wind speed available at a particular site is the first criterion to be considered in site selection. During the site identification process, the measuring instruments described in the previous section are installed at the site. The annual average wind speed is calculated according to the equation

$$\bar{v} = \frac{1}{t_2 - t_1} \int_{t_1}^{t_2} v \, dt \tag{2.2}$$

where $\bar{v}$ is the annual average wind speed (m/s), v is the instantaneous wind speed (m/s), and $t_2 - t_1$ is the duration of one year (8760 hrs).

In the case of a digital data logger recording wind speed data at regular intervals, the average wind speed can be calculated as

$$\bar{v} = \frac{1}{n} \sum_{i=1}^{n} v_i$$

where v_i is the wind speed at the ith observation and n is the number of observations.

At any given site, the wind speed varies with the height from the ground level. It is generally not possible to install measuring instruments at all heights, but an empirical formula can be used to find the mean wind speed at a certain height using the observed mean wind speed at 10 m:

$$\bar{v}_H = \bar{v}_{10} \left(\frac{H}{10}\right)^x \tag{2.3}$$

where $\bar{v}_H$ is the annual average wind speed at height H (m/s), $\bar{v}_{10}$ is the annual average wind speed at 10 m (m/s), and x is an exponent that depends on the roughness of the ground. The values of x are given in Table 2.1.

The measuring instruments record the wind speed continuously against time. If the data are collected throughout a year, the

Table 2.1 Values of the boundary layer exponent x for varying ground roughness

Description of land	Exponent x
Smooth land with very few obstacles (e.g., sea, coast, desert, snow)	0.10–0.13
Moderately rough, (e.g., agricultural fields with very few trees, grasslands, rural areas)	0.13–0.20
Land with uniformly distributed obstacles 10–15 m high (e.g., forests, small towns, agricultural fields with tree plantations)	0.20–0.28
Land with big and non-uniform obstacles (e.g., big cities, plateaus)	0.28–0.40

resulting chart would look like a long wavy line, as shown in Fig. 2.3.

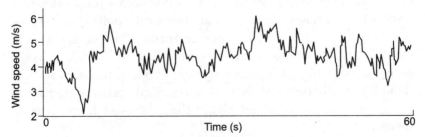

Fig. 2.3 Typical anemometer recording of wind speed versus time

It is clear that this plot is too wavy and irregular for us to obtain any useful information from it. The next step is to obtain, from Fig. 2.3, the plot of wind speed v versus the total time during a year for which the wind speed is v (Fig. 2.4); called the *wind speed distribution curve*. Of course, the period of time for which the wind speed assumes an exact value is infinitely small. So, the vertical axis actually gives the annual duration for which the wind speed falls within certain limits, for instance 0.5 m/s below and above v. The y-axis of the wind speed distribution curve should be given in hours per annum per metres per second. Thus the integral of the function (or the area under the curve) will always be 8760 hrs, corresponding to the number of hours in a year.

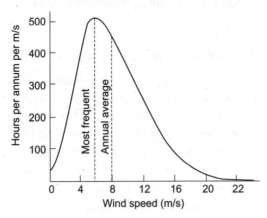

Fig. 2.4 Plot with wind speed on the x-axis and the duration in a year for which wind assumes that speed on the y-axis

It is easy to plot the *energy distribution curve* for a site—the energy available at a particular wind speed is the power contained in the wind (proportional to v^3, the value of the proportionality constant is unimportant for our purpose) multiplied by the duration for which wind blows at that speed. This curve, shown in Fig. 2.5, gives the value of the wind speed at which the maximum energy is available—it is the wind speed at which the wind turbine should normally be rated. Note that the wind speed for maximum energy is different from and higher than the most frequent wind speed.

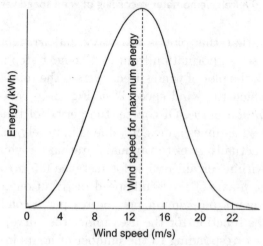

Fig. 2.5 The energy distribution curve

2.2.1 Statistical Wind Speed Distributions

In certain cases, the total data of wind speeds against time over a year may not be available, but the yearly average wind speed may be known. In such cases, the wind speed distribution curves can be obtained approximately from the magnitude of the average wind speed, by using a standard statistical distribution function, such as the Rayleigh distribution function. It is observed that the wind speed distributions of different sites have certain similarities and can be approximated by the Rayleigh distribution function.

The distribution function is given by

$$t = 8760 \frac{\pi}{2} \frac{v}{\bar{v}^2} \exp\left(\frac{-\pi v^2}{4\bar{v}^2}\right) \tag{2.4}$$

where t is the time (hours per year), v is the wind speed (m/s), and $\bar{v}$ is the annual average wind speed (m/s). Equation (2.4) predicts the total number of hours per year for which wind will blow at speed v at a site with mean wind speed $\bar{v}$. It can be shown analytically that for a Rayleigh distribution, the most frequent wind speed occurs at $v_{mf} = 0.8\bar{v}$ and the maximum energy is available at $1.6\ \bar{v}$. These relations give a very quick method of finding the wind speed at which the maximum energy is available, that is, the speed at which a wind turbine should be rated. It should be noted, however, that the Rayleigh distribution becomes inappropriate at wind speeds below 10 mph, and therefore should not be used for sites where the mean annual wind speed is below 10 mph.

A more general distribution function is required to obtain a better approximation for wind speed distribution on a daily or a still shorter time scale. In such cases, one may apply the Weibull distribution, given by

$$f(v) = \left(\frac{k}{c}\right)\left(\frac{v}{c}\right)^{k-1} \exp\left[-\left(\frac{v}{c}\right)^k\right] \tag{2.5}$$

where c is a scale factor often taken equal to the mean wind speed calculated at hub height, k is the shape factor, ranging between 1.8 and 2.3, and

$$f(v) = \frac{\text{fraction of time the wind speed is between } v \text{ and } (v + \Delta v)}{\Delta v}$$

The curves calculated using Eqn (2.5) with $k = 1.8$ and $k = 2.3$ are shown in Fig. 2.6(a). The value of k is chosen to fit the actual curve in the best way. It may be noted that for $k = 2$, the Weibull distribution reduces to the Rayleigh distribution when converted into the proper units.

The dependence of the distribution function on the choice of the scale factor c is shown in Fig. 2.6(b). For greater values of c, the curve shifts to the right, to higher wind speeds, which implies that high wind speeds are experienced for a greater number of days. As stated earlier, a good choice of c for a particular site is the annual average wind speed $\bar{v}$.

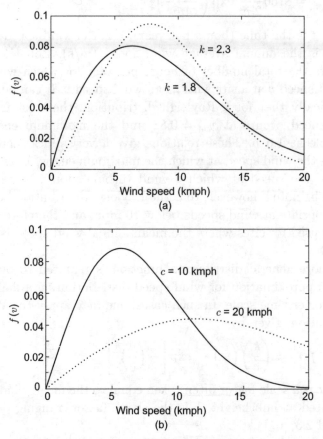

Fig. 2.6 The Weibull distribution for (a) $k = 1.8$ and 2.3, with c set at mean wind speed $\bar{v} = 10$ kmph, and (b) $c = 10$ and 20 kmph, with k set at 2

We can get an idea about the energy potential of a site from the root mean cube (rmc) speed, which is given by

$$V_{\mathrm{rmc}} = \left(\frac{1}{8760} \int_0^\infty f(v)v^3 dv \right)^{1/3} \qquad (2.6)$$

In terms of discrete observations of anemometer readings, the rmc wind speed can be calculated using the formula

$$\overline{v} = \left(\frac{\sum\limits_{j=1}^{N} V_j^3}{N} \right)^{1/3} \tag{2.7}$$

where V_j is the wind speed at the jth observation and N is the number of wind speed observations. The value of V_{rmc} is very useful in estimating the annual average power of a site, given by

$$P_{\text{rmc}} \approx \frac{1}{4} \, \rho \, V_{\text{rmc}}^3 \;\; (\text{W/m}^2)$$

We thus obtain, either from direct measurement or by using this statistical distribution formula, the wind speed distribution curve shown in Fig. 2.4.

Example 2.1 The annual average wind velocity at a height of 10 m over a flat terrain is 6 m/s. The boundary layer exponent is $x = 0.13$. Find the annual average power density (W/m^2) in the wind at a height of 50 m. Assume the Rayleigh distribution as an approximation to the wind velocity–duration distribution over the terrain and 1.225 kg/m^3 as the density of air.

Solution   The Rayleigh distribution is a special case of the Weibull function when the shape parameter k in Eqn (2.5) is equal to 2. Hence the Rayleigh distribution function is given by

$$f(v) = \frac{2v}{c^2} e^{-(v/c)^2} \tag{2.8}$$

The average wind speed is

$$\overline{v} = \int_0^\infty v \, f(v) dv \tag{2.9}$$

Substituting the expression for $f(v)$ in Eqn (2.9) and carrying out the integration give the average wind speed

$$\overline{v} = c\Gamma(1.5) \tag{2.10}$$

where $\Gamma(1.5)$ is the gamma function. Using tables for the gamma function used in Eqn (2.10) gives

$$c = 1.12\bar{v} \tag{2.11}$$

The annual average wind velocity at a height of 50 m is

$$\bar{v}_{50} = \bar{v}_{10}\left(\frac{50}{10}\right)^{0.13} = 7.52 \text{ m/s} \tag{2.12}$$

At this wind velocity, the scale factor from Eqn (2.11) is

$$c = 1.12 \times 7.52 = 8.42 \text{ m/s} \tag{2.13}$$

The power in the wind is

$$P_w = \frac{1}{2}\rho A v_{50}^3 \text{ (W)} \tag{2.14}$$

The average power in the wind is given by

$$P_{\text{avg}} = \frac{1}{2}\rho A \int_0^\infty v^3 f(v)dv \text{ (W)} \tag{2.15}$$

The use of Eqn (2.8) in Eqn (2.15) yields

$$P_{\text{avg}} = \frac{1}{2}\rho A \int_0^\infty \frac{2v^4}{c^2}e^{-(v/c)^2}dv \tag{2.16}$$

Introducing the change in variable $x = (v/c)^2$ in Eqn (2.16) yields

$$P_{\text{avg}} = \frac{1}{2}\rho A c^3 \Gamma\left(1 + \frac{3}{2}\right) \tag{2.17}$$

Dividing Eqn (2.17) by A, consulting the table for gamma functions, and substituting the required values yield the following average power density:

$$P_{\text{avg}} = \frac{1}{2} \times 1.225 \times 8.42^3 \times 1.3$$

$$= 475.3 \text{ W/m}^2$$

2.3   Site and Turbine Selection

Site selection involves not only the choice of the geographical location for a wind turbine or a wind farm, but also the model of the turbine that is best suited to a particular site.

For the final selection process, that is, while choosing the wind turbine that is best suited for a particular site, a modification of the curve shown in Fig. 2.4 is necessary. At this stage, we plot the speed–duration curve—the graph of v versus the total duration for which the wind speed exceeds or equals v (Fig. 2.7). Naturally, the largest coordinate on the y-axis is the number of hours in a year (8760), when the wind speed exceeds zero. If the wind speed is measured using a digital recorder with data logging facility, the wind speed distribution and duration curves can be obtained directly or generated by a computer later using the stored data.

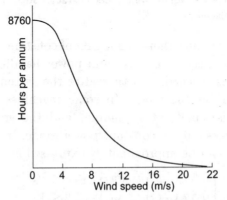

Fig. 2.7 The wind speed–duration curve: plot with wind speed along the x-axis and the duration for which the wind speed equals or exceeds that speed along the y-axis

The productivity of any wind generator at a particular site depends on the characteristics of the site (given by Fig. 2.7) and those of the wind machine. The latter are given as the power versus wind speed characteristics (such as that shown in Fig. 2.8), which are generally available for all commercially produced wind machines.

Every wind turbine model has a specific cut-in speed, a rated speed, a furling speed, and power versus wind speed characteristics within the wind speed range between the cut-in speed and the furling speed. At the cut-in speed the wind generator starts generating power. As the wind speed increases, the power output increases in proportion with the power contained in the wind. After the rated speed is reached, the speed-regulating mechanism

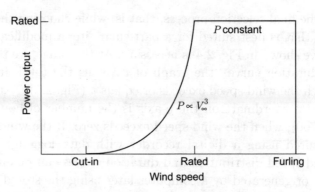

Fig. 2.8 Typical power versus wind speed characteristics of constant-speed wind machines

comes into action, and there is a region of constant speed. Beyond a certain wind speed, the maximum power handling capacity of the generator is reached, and thereafter the system works in the constant-power output mode. In some machines the constant-speed region is small (or negligible) and the speed-regulating mechanism works only in constant-power mode. In such cases the characteristics can be approximately expressed as

$$P(v) \approx \begin{cases} 0.5\eta_v C_p \rho A v^3 \text{ for } V_c \leq v < V_r \\ 0.5\eta_v C_p \rho A V_r^3 \text{ for } V_r \leq v < V_f \end{cases} \tag{2.18}$$

where V_c is the cut-in speed, V_r is the rated speed, V_f is the furling speed, η_v is the efficiency of generator and mechanical transmission, C_p is the wind turbine coefficient of performance, ρ is the density of air, A is the blade swept area, and v is the wind speed. At the furling wind speed, the plant is shut down to avoid damage.

From Figs 2.8 and 2.7, the wind generator's characteristics, as weighted by the site's wind speed–duration curve, yield the power–duration characteristics. For each value of wind speed shown in Fig. 2.7, the corresponding value of the output power is obtained from Fig. 2.8. The typical output power–duration curve for a wind turbine is shown in Fig. 2.9. To illustrate, we have also shown the wind power–duration curve (obtained by the relation $0.5\eta_v C_p \rho A v^3$), so that the energy loss due to cut-in and furling becomes clear.

The area under the output power–duration curve measures the energy output of a particular machine at a given site. By plotting

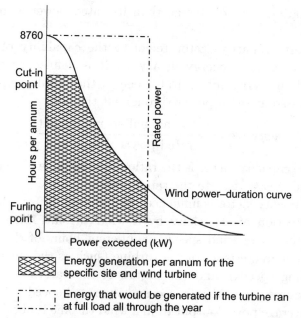

Fig. 2.9 The output power–duration characteristics of a wind generator at a given site

similar curves for different machines at a particular site, one can choose the appropriate machine. One generally chooses the model that gives the maximum output for a specific rated power at a particular site.

2.4 Capacity Factor

Wind power plants differ in a variety of ways from power plants that burn fuel. In spite of the downtime in a year, a coal plant can be run day and night at almost its rated capacity during any season of the year. In contrast, the wind speed varies with the time of the day and with the season. At times the wind speed may even be insufficient to drive the turbine. Consequently, a wind turbine cannot operate 24 hrs a day, 365 days a year at full power. A wind farm generally runs 65–80 % of the time in a year with variation in output power. Because wind farms get paid for the total energy production, the annual energy output is a more relevant measure for evaluating a wind turbine than its rated power at a certain speed.

The term capacity factor refers to the capability of a wind turbine to produce energy in a year. It is defined as the ratio of the actual energy output to the energy that would be produced if it operated at rated power throughout the year.

$$\text{Capacity factor} = \frac{\text{annual energy output}}{\text{rated power} \times \text{time in a year}}$$

Thus, the capacity factor is the ratio of the average output power, computed over a year, to the rated power.

The capacity factor is influenced by the same factors that affect the production of electricity by a wind turbine. These factors include the mean wind speeds at different hours of the day, the type and characteristics of the turbine (such as the cut-in, rated, and furling speeds), and the nature of the variation of output power between the cut-in speed and the rated speed.

The average power output from a wind turbine is obtained from the product of the power produced at each wind speed and the fraction of time for which this speed prevails, integrated over all possible wind speeds. In terms of the probability distribution $f(v)$ of the wind speed–duration curve, the average power is

$$P_{\text{avg}} = \int_0^\infty P(v) f(v)\, dv \tag{2.19}$$

When the total data of wind speeds over a year are not available but the yearly average wind speed at a particular site is known, the Weibull distribution function given by Eqn (2.5) may be used instead of $f(v)$. The equation describing the variation in output power between the cut-in and furling speeds is given by Eqn (2.18). Substituting $P(v)$ from Eqn (2.18) into Eqn (2.19) and dividing the result by the rated power, we get the capacity factor (CF):

$$\text{CF} = \frac{1}{V_r^3} \int_{V_c}^{V_r} v^3 f(v)\, dv + \int_{V_r}^{V_f} f(v)\, dv$$

If a closed form expression cannot be obtained after integration, numerical integration will be required. Capacity factors of successful wind farms usually lie in the range 0.20–0.35, which is quite low compared to the capacity factors of 0.5–0.75 for fossil fuel plants.

Summary

This chapter introduces the various techniques of wind speed measurement. Of these techniques, the Robinson cup anemometer is most popular, because the output is obtained in the form of an electrical signal. Data loggers can then be used to keep a record of the wind speeds occurring at short time intervals, which can later be loaded into a computer for further analysis.

The chapter also introduces the techniques of analysing the mass of data that is obtained from direct measurement at any given site. Statistical techniques to obtain the approximate wind speed distribution curve at any site are presented.

A major problem faced by wind system planners is to choose a wind turbine that will be suit the specific characteristics of a particular site. The technique to achieve this matching has also been introduced.

Problems

Refer to Fig. 2.10 for solving the following problems.

1. Replot the wind speed distribution curve for Chandipur with the x-axis expressed in m/s and the y-axis expressed in hours per annum per m/s.

2. Plot the energy distribution curve for the four sites and obtain the wind speeds at which the maximum amount of energy is available.

3. For the wind speed distribution curve for Chandipur, obtain the annual average wind speed. Draw the energy distribution curve using the Rayleigh distribution. Compare it with the actual distribution for the site.

4. Estimate the parameters k and c of the Weibull distribution function for Puri. Plot the curve and compare it with the actual wind speed distribution for the site.

5. Assuming a boundary layer exponent of 0.14, obtain the wind speed distribution at a hub height of 18 m for the site of Gopalpur.

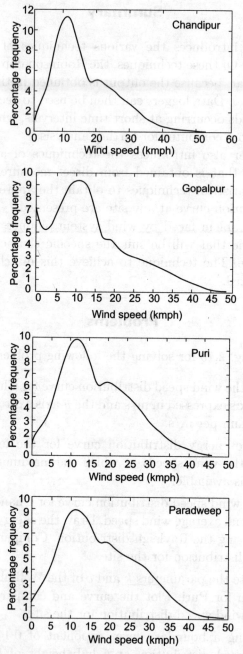

Fig. 2.10   The measured wind speed distribution curves for four sites in the state of Orissa, at a height of 10 m (Mani 1990)

6. Draw the wind speed duration curves for the four sites.

7. Table 2.2 gives the wind data for a site in terms of the percentage of time over a year for different speed groups. Calculate the annual average power in the wind passing normally through the swept area of a turbine of diameter 30 m. Take $1.225 \, \text{kg/m}^3$ as the density of air.

Table 2.2

Speed group (m/s)	$0 < v \leq 3$	$3 < v \leq 6$	$6 < v \leq 9$	$9 < v \leq 12$	$12 < v \leq 16$	$16 < v \leq 20$
Percentage of time	12.36	28.25	29.37	18.96	9.31	1.67

8. The wind speed distribution over a year at a site is described by the two-parameter Weibull density function in which the shape parameter k is 2.6 while the scale parameter is 1.1 times the mean speed $\bar{v}$. If the rated wind speed to cut-in wind speed ratio is 2.5, find the value of the rated wind speed in terms of the mean wind speed $\bar{v}$ for which the energy extracted per year will be maximum.

9. The specifications of a few commercially available wind turbines are given in Table 2.3. Using the power versus wind speed characteristics shown in Fig. 2.8, obtain the capacity factor and find out which windmill is most suitable for each of the four sites.

10. A wind turbine rated at 100 kW has a rated wind speed of 10.5 m/s, a cut-in speed of 4.5 m/s, and a furling speed of 22 m/s. The wind speed frequency distribution over a year is given by a Weibull distribution having the shape parameter $k = 5$ and the scale parameter $c = 7 \, \text{m/s}$. Determine the capacity factor and the yearly energy production.

11. For the wind frequency distribution of Problem 10, find the optimum value of the rated wind speed for which the energy production per year will be maximum.

12. The coefficient of performance C_p versus tip speed ratio λ for a fixed-speed (42 rpm) turbine of diameter 30 m is given by

$$C_p = C_{p,\text{max}} \frac{\lambda}{\lambda_{\text{opt}}} \left(2 - \frac{\lambda}{\lambda_{\text{opt}}} \right)$$

Table 2.3 Specifications of a few commercially available wind turbines

Windmill	Rated power (kW)	Cut-in wind speed (m/s)	Rated wind speed (m/s)	Furling wind speed (m/s)	Hub height (m)
1	75	4.5	8.9	22.3	21.3
2	3000	6.3	11.8	24	100
3	50	5.4	9.8	19.7	18.3
4	2500	6.3	12.2	26.8	60.9
5	2000	6	13	21	80
6	100	5	13	20	18.3
7	15	4.5	9.5	25	15
8	15	4	10.3	27	18.3
9	4	3.6	12	17.8	24.4
10	0.75	6	7	14	10
11	5.5	4	14	25.3	18.7
12	3.5	3	11	20	18.7
13	2.2	4	11	12	10
14	10	5	12	12.1	14

C_p reaches the maximum value $C_{p,\max} = 0.38$ at $\lambda = \lambda_{\text{opt}}$ (= 6). Plot the shaft power output of the turbine at standard conditions (0° and 101.3 kPa), which yield a density of 1.293 kg/m^3 for dry air.

3 Basics of Induction and Synchronous Machines

The generation schemes for wind electrical conversion systems depend primarily on the type of output required as well as the mode of operation of the turbine. In present-day practice, two types of generators generally find application in wind power plants: the synchronous generator and the induction generator. Synchronous generators may have dc field excitation or a permanent magnet field. Systems using line-frequency excited alternators and ac commutator generators have been suggested for constant-frequency output from an aeroturbine operated in the variable-speed mode, but they are not preferred over synchronous and induction generators.

The theory of induction and synchronous machines is found in all text books on electrical machines. However, we will present some of their basic aspects here to facilitate a proper understanding of the operation of these machines as electrical generators. In this chapter, we will briefly discuss the operation, circuit models, and characteristic features of these machines.

3.1 The Induction Machine

Besides being commonly used as drives in the industry, three-phase induction machines have earned much favour as wind generators because of qualities such as ruggedness, reliability, and manufacturing simplicity. They constitute the largest segment in the wind power industry today. Two types of three-phase induction machines are used: the squirrel cage type and the wound rotor (slip-ring) type. In the kilowatt range, the former is favoured, whereas in the megawatt range, the latter is favoured. Their principles of operation are basically the same; they differ only with respect to their application.

3.1.1 Constructional Features and Rotating Magnetic Field

A three-phase induction motor has three identical windings that are symmetrically distributed around the inner surface of a laminated cylindrical shell called the *stator*. The laminated rotor inside carries either a winding consisting of bars connected to two shorting rings at both ends, or three-phase balanced windings connected to three slip rings, which in turn can be connected to external circuits through brushes touching the slip rings. The former is known as a *squirrel cage rotor* and the latter is known as a *wound rotor*. The squirrel cage rotor can be adapted to any number of stator poles while the windings of the slip-ring rotor must be wound for the same number of poles as for the stator.

When the balanced stator windings, wound for p pole pairs and displaced in space by 120 electrical degrees, are excited by a balanced three-phase sinusoidal supply, the current in the three phases of the stator winding (also displaced in time-phase by 120 electrical degrees) creates a rotating magnetic field of constant strength. The sinusoidally distributed magnetic field moves around the air gap at a speed given by

$$n_s = \frac{f}{p} \ (\text{rps}) \tag{3.1}$$

which is known as the synchronous speed of the induction machine.

The rotating air-gap field induces electromotive force (emf) in the rotor windings, and these emfs in turn produce currents in the short-circuited rotor bars or windings. The interaction between these rotor currents and the air-gap flux creates a torque which, by Lenz's law, acts in the direction of the rotating field. For the production of torque, a speed difference between the rotor and the air-gap field is necessary, so that the rotor flux linkage can change. A very important variable in the performance analysis of an induction machine is its per-unit slip s, which is expressed as

$$s \triangleq \frac{n_s - n_r}{n_s} \tag{3.2}$$

where n_r is the rotor speed in rps. For motoring operation, $n_r < n_s$, i.e., $0 < s < 1$.

As the relative speed between the rotor and the air-gap field is sn_s $(= n_s - n_r)$, current will be induced in the rotor at the slip frequency sf. The slip frequency rotor currents create a rotor magnetomotive force (mmf), the space fundamental component of which moves past the rotor at the slip speed $n_s - n_r$. The rotor speed being n_r, the rotor mmf and the air-gap field are stationary with respect to each other. The torque magnitude depends on the *space angle* δ, also called the *torque angle*, between these two waves in addition to their magnitudes. The rotor mmf lags behind the air-gap field. It can be shown that this torque angle δ equals $90° + \phi_2$, ϕ_2 being the rotor power-factor angle at the slip frequency sf.

3.1.2 Steady-state Equivalent Circuit Model

It is desirable to obtain an equivalent circuit of the induction machine to facilitate the analysis and computation of its performance. For balanced operation, a per-phase circuit model will suffice, irrespective of the delta or star connection of the stator windings.

The stator circuit model

The resultant air-gap flux ϕ_m is set up by the combined action of the stator and the rotor mmfs. This synchronously rotating flux induces a counter emf E_1 in the stator phase winding. The stator terminal voltage V_1 differs from this counter emf E_1 by the stator leakage impedance drop. The stator emf phasor equation is then

$$\mathbf{V}_1 = \mathbf{E}_1 + \mathbf{I}_1(R_s + jX_{ls}) \tag{3.3}$$

where

$$E_1 = \sqrt{2}\pi f k_{w_1} T_1 \phi_m \tag{3.4}$$

The circuit representation of Eqn (3.3) is shown in Fig. 3.1. Just as in the case of a transformer primary winding, the stator winding current $\mathbf{I}_1$ can be resolved into an exciting component $\mathbf{I}_0$ and a compensating load component $\mathbf{I}_2'$. The load component current $\mathbf{I}_2'$ counteracts the rotor mmf, thereby demanding power from the source. The exciting component $\mathbf{I}_0$ can be resolved into a core-loss component $\mathbf{I}_c$, in phase with the stator induced emf $\mathbf{E}_1$ and a magnetizing component $\mathbf{I}_m$, lagging behind the induced emf $\mathbf{E}_1$ by 90°. It is this magnetizing component which sets up the air-gap flux ϕ_m.

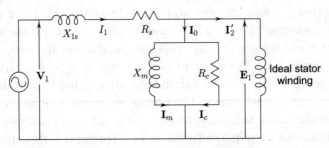

Fig. 3.1   Per-phase stator circuit model of a three-phase induction motor

The rotor circuit model

The air-gap flux wave induces a slip frequency rotor voltage E_{2s} given by

$$E_{2s} = \sqrt{2}\pi s f k_{w_2} T_2 \phi_m \qquad (3.5)$$

With respect to Fig. 3.2(a), the rotor current per phase may be expressed as

$$\mathbf{I}_{2s} = \frac{E_{2s}}{R_r + jsX_{\mathrm{lr}}} \qquad (3.6)$$

It may be noted here that the current I_{2s} flows at slip frequency. Combining Eqns (3.4), (3.5), and (3.6),

$$\mathbf{I}_{2s} = \frac{s\mathbf{E}_1}{m_1(R_r + jsX_{\mathrm{lr}})} \qquad (3.7)$$

where m_1 is the ratio of the effective stator turns ($k_{w_1} T_1$) to the effective rotor turns ($k_{w_2} T_2$). Equation (3.7) can also be rewritten in the form

$$\frac{\mathbf{I}_{2s}}{m_1} = \frac{\mathbf{E}_1}{R_r'/s + jX_{\mathrm{lr}}'} \qquad (3.8)$$

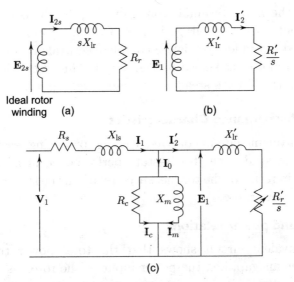

Fig. 3.2 Steady-state per-phase circuit models of a three-phase induction motor: (a) rotor circuit model at slip frequency, (b) rotor circuit model at stator frequency, (c) complete stator-referred T-form equivalent circuit

where the stator-referred rotor resistance $R'_r = m_1^2 R_r$ and the stator-referred rotor leakage reactance $X'_{1r} = m_1^2 X_{1r}$. From the mmf balance between the ideal stator circuit and the rotor circuit (vide Figs 3.1 and 3.2), the stator-referred rotor current $I'_2 = I_{2s}/m_1$. We can make a clear distinction between Eqns (3.7) and (3.8). The former holds good at the slip frequency and the latter at the line frequency, as $\mathbf{I}'_2$ arises out of the line-frequency voltage E_1 in a circuit with impedance $R'_r/s + jX'_{1r}$.

Equation (3.8) suggests that the actual rotor circuit in Fig. 3.2(a) may be replaced by the stator-referred equivalent circuit shown in Fig. 3.2(b) at the line frequency. In both the circuits, the current maintains the same phase relation with respect to the corresponding input voltage. The power-factor angle of the rotor circuit is given by

$$\phi_2 = \tan^{-1} \frac{sX_{1r}}{R_r}$$

$$= \tan^{-1} \frac{X'_{1r}}{R'_r/s} \tag{3.9}$$

Since the input terminal voltage in Fig. 3.2(b) is the same as the output terminal voltage in Fig. 3.1, the two circuits may be connected in tandem to yield a complete equivalent circuit for the induction motor, as shown in Fig. 3.2(c). This circuit is generally known as the T-form steady-state equivalent circuit.

3.1.3 Performance Characteristics

The equivalent circuit derived in the preceding sections can be used to study the steady-state performance of an induction machine in terms of the torque–speed relation, the stator and rotor currents, power, etc.

Torque and power relations

The equivalent circuit shows that the total power transferred across the air gap, i.e., the power input to the rotor, is

$$P_{ag} = 3E_1 I'_2 \cos \phi_2$$
$$= 3 \frac{I'^2_2 R'_r}{s}$$

and by Eqn (3.8),

$$P_{ag} = \frac{3E_1^2 s R'_r}{R'^2_r + s^2 X'^2_{lr}} \tag{3.10}$$

Under actual operating conditions, the electrical power consumed in the rotor circuit is the rotor copper loss, which is given by

$$P_{Cu2} = 3I_2^2 R_r$$
$$= 3I'^2_2 R'_r$$
$$= s P_{ag} \tag{3.11}$$

The effective mechanical power is then given as

$$P_m = P_{ag} - P_{Cu2}$$
$$= 3I'^2_2 R_r \left(\frac{1-s}{s} \right)$$
$$= (1-s) P_{ag} \tag{3.12}$$

The power expressions in Eqns (3.11) and (3.12) show that

$$P_{ag} : P_{Cu2} : P_m = 1 : s : (1-s) \tag{3.13}$$

The electrical power component, which is the slip multiplied by the total power delivered to the rotor, is called the *slip power* of the system. If the power components in Eqns (3.11) and (3.12) are to be emphasized in the equivalent circuit of Fig. 3.2(c), the effective rotor resistance must be split into two components as shown in Fig. 3.3(a). The power dissipation in the slip-dependent resistance $[(1-s)/s]R_r'$ is the developed mechanical power per phase. Figure 3.3(b) presents the corresponding phasor diagram.

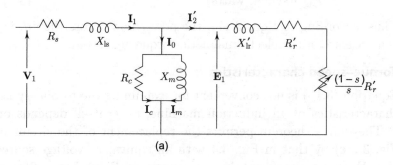

(a)

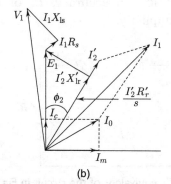

(b)

Fig. 3.3 (a) Stator-referred circuit model; (b) phasor diagram

The electromagnetic torque developed by an induction motor is

$$T_e = \frac{P_m}{2\pi n_r} \ (\mathrm{N\,m}) \tag{3.14}$$

Using the expressions (3.2) and (3.12) in Eqn (3.14),

$$T_e = 3\frac{I_2'^2 R_r'}{s}\frac{1}{2\pi n_s} \ (\mathrm{N\,m}) \tag{3.15}$$

$$= \frac{P_{\text{ag}}}{2\pi n_s} \quad (\text{N m}) \tag{3.16}$$

Since, for a given supply frequency, the synchronous speed $2\pi n_s$ is constant, T_e is called the torque in synchronous watts and is frequently expressed in the following form:

$$T_e = 3\frac{I_2'^2 R_r'}{s}$$

$$= P_{\text{ag}} \quad (\text{synch. watts}) \tag{3.17}$$

Thus, according to Eqn (3.17), the power flow across the air gap is a measure of the electromagnetic torque generated.

Torque–speed characteristics

Equation (3.15) is not convenient for examining the torque–speed characteristics of an induction machine, as I_2' itself depends on s. Thevenin's theorem permits the replacement of the circuit in Fig. 3.2(c) by that in Fig. 3.4 with an equivalent voltage source V_{Th}. With no significant loss in accuracy, R_c of Fig. 3.2(c) is ignored for the sake of simplicity.

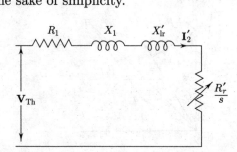

Fig. 3.4   Thevenin's equivalent of the circuit in Fig 3.2(c)

$$\mathbf{V}_{\text{Th}} = \mathbf{V}_1 \frac{jX_m}{R_s + j(X_{\text{ls}} + X_m)} \tag{3.18}$$

$$= k\mathbf{V}_1 \tag{3.19}$$

The new impedance elements in Thevenin's equivalent circuit are

$$R_1 + jX_1 = \frac{(R_s + jX_{\text{ls}})(jX_m)}{R_s + j(X_{\text{ls}} + X_m)} \tag{3.20}$$

Evaluating the rotor current from Thevenin's equivalent circuit and inserting it into the torque equation (3.15), the following form of the induction motor torque is obtained:

$$T_e = \frac{3}{2\pi n_s} \frac{k^2 V_1^2 R_r'/s}{(R_1 + R_r'/s)^2 + (X_1 + X_{lr}')^2} \tag{3.21}$$

If an external resistance R_x is inserted into the rotor circuit, the torque expression becomes

$$T_e = \frac{3}{2\pi n_s} \frac{k^2 V_1^2 (R_r' + R_x')/s}{[R_1 + (R_r' + R_x')/s]^2 + (X_1 + X_{lr}')^2} \tag{3.22}$$

where R_x' is the stator-referred value of the rotor external resistance R_x.

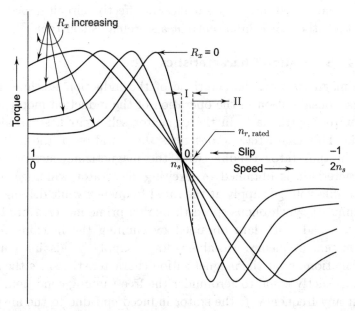

Fig. 3.5 The torque–speed characteristics of an induction motor showing the effect of rotor resistance variation

The general shapes of the torque–speed (torque–slip) curves for varying values of the external rotor resistance under constant machine terminal voltage are shown in Fig. 3.5. Note that with increase in the value of the rotor circuit resistance, the slip increases for the same value of the torque. A negative value of the slip implies running the machine above synchronous speed in the direction of the rotating field. As the torque direction is simultaneously reversed (opposite to the direction of the rotating field), the machine has to be driven by a source of mechanical power to counteract the opposing torque. In the process the machine acts as a generator feeding power to the source. For $s > 1$, the machine runs in a direction opposite to that of the rotating field and the internal torque. In order to sustain this condition, the machine should also be driven by a mechanical power source. This mode of operating the induction machine is known as *plugging*, and is equivalent to an electrical braking method.

Equation (3.22) reveals that, for a constant input voltage, if the ratio of the total rotor circuit resistance to the slip, i.e., $(R_r + R_x)/s$, remains constant, the developed torque T_e would also remain constant. As this makes the effective circuit impedance constant, the stator input current also remains constant.

3.1.4 Saturation Characteristics

The magnetization characteristic of the induction machine is a prime consideration in the operation of the induction motor as a generator, particularly in the capacitor self-excited, stand-alone mode. This characteristic describes the variation of the induced emf in the stator winding with the magnetizing current. The characteristic is obtained by exciting the stator windings from a variable-voltage supply at the rated frequency while driving the machine at synchronous speed through a prime mover. Note that the 'no-load test' data obtained by running the machine from a constant-frequency, variable-voltage supply yields inaccurate information about the magnetization characteristic, since the slip is not exactly equal to zero under the free running condition.

At any frequency f, the stator-induced emf due to the air-gap flux is given by

$$E_1 = \sqrt{2}\pi f \phi_m T_1 k_{w_1}$$

i.e.,

$$\frac{E_1}{F} = \sqrt{2}\pi f_b \phi_m T_1 k_w \tag{3.23}$$

where F is the ratio of the actual frequency f to the base frequency f_b.

Figure 3.6(a) is a typical plot of the magnetization characteristics relating the sine rms induced voltage E_1/F to the sine rms magnetizing current I_m at the base frequency (i.e., $F = 1$). It is observed that the relationship is not linear as may be expected from the equivalent circuit of Fig. 3.1 with constant X_m. Due to saturation of the stator core, X_m does not remain constant, as the magnitude of the air-gap flux, and hence E_1/F, increases. Equation (3.23) suggests that an actual measurement of the saturation characteristic at the base frequency is sufficient to determine the saturation level at any other frequency for a given air-gap voltage. The modelling of the magnetization characteristic depends on the method of analysis adopted for the machine. It may be modelled in a piecewise manner by a polynomial of degree n. For example,

$$\frac{E_1}{F} = a_0 + a_1 I_m + a_2 I_m^2 + \cdots + a_n I_m^n \tag{3.24}$$

for $I_{m1} \leq I_m \leq I_{m2}$ (Fig. 3.6). An alternative way of depicting the magnetization characteristic is to plot from the curve in Fig. 3.6(a) the variation of E_1/F with the magnetizing reactance as shown in Fig. 3.6(b). The saturated portion of this characteristic can be closely represented by a linear equation of the form

$$\frac{E_1}{F} = k_1 - k_2 X_m \tag{3.25}$$

where the constants k_1 and k_2 depend on the design of the machine.

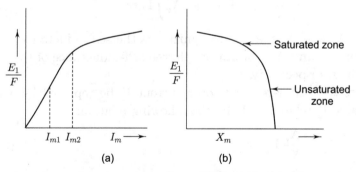

Fig. 3.6 (a) Magnetization characteristic of an induction machine at synchronous speed; (b) variation of the effective magnetizing reactance with flux level at synchronous speed

3.1.5 Modified Equivalent Circuits

The inverse-Γ model

The T-form steady-state equivalent circuit shown in Fig. 3.2(c) is adequate in many situations. However, it can be transformed into other simpler models without any loss of accuracy. In fact, such circuits are found to be more convenient for analysing the behaviour of the induction machine in certain situations. One such circuit model of the machine is known as the inverse-Γ equivalent circuit.

No significant error is introduced in the prediction of the performance of the machine if the shunt resistance, representing the iron loss in the stator, is ignored in favour of simplicity of manipulating the equations.

Applying Kirchhoff's circuit laws to the circuit of Fig. 3.7(b), the following equations can be written:

$$\mathbf{V}_1 = \mathbf{I}_1(R_s + jX_{ls}) + j(\mathbf{I}_1 - \mathbf{I}_2')X_m$$
$$0 = \mathbf{I}_2'\left(\frac{R_r'}{s} + jX_{lr}'\right) - j(\mathbf{I}_1 - \mathbf{I}_2')X_m \tag{3.26}$$

Upon rearranging the terms and putting $X_s = X_{ls} + X_m$ and $X_r' = X_{lr}' + X_m$, the above equations take the forms

$$\mathbf{V}_1 = \mathbf{I}_1(R_s + jX_s) - j\mathbf{I}_2'X_m$$
$$0 = -j\mathbf{I}_1X_m + \left(\frac{R_r'}{s} + jX_r'\right)\mathbf{I}_2' \tag{3.27}$$

Here X_s and X_r' may be recognized as the self-inductance of the stator winding and the stator-referred self-inductance of the rotor winding, respectively.

Let the stator-referred rotor current $\mathbf{I}_2'$ be replaced by a new variable $\mathbf{I}_2^e$ related to $\mathbf{I}_2'$ by the following relation:

$$\mathbf{I}_2' = \frac{X_r'}{X_m}\mathbf{I}_2^e \tag{3.28}$$

Using Eqn (3.28) in Eqn (3.27) and rearranging the various terms,

$$\mathbf{V}_1 = \mathbf{I}_1(R_s + jX_\sigma) + (\mathbf{I}_1 - \mathbf{I}_2^e)\frac{X_m^2}{X_r'}$$

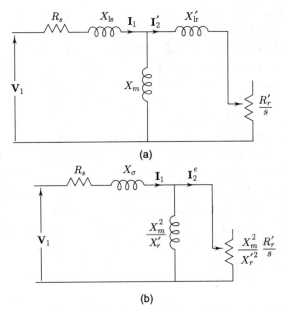

Fig. 3.7 Stator-referred, steady-state equivalent circuits: (a) T-form,
(b) inverse-Γ-form

$$0 = -(\mathbf{I}_1 - \mathbf{I}_2^e)\frac{X_m^2}{X_r'} + \mathbf{I}_2^e\frac{X_m^2}{X_r'^2}\frac{R_r'}{s} \tag{3.29}$$

where

$$X_\sigma = X_s - \frac{X_m^2}{X_r'}$$

$$= X_{ls} + \frac{X_m X_{lr}'}{X_m + X_{lr}'} \tag{3.30}$$

In this result, X_σ is the familiar expression for the transient
reactance of the induction machine. The information contained
in Eqns (3.28) and (3.29) may be expressed by the equivalent
circuit shown in Fig. 3.7(b). The loss in the slip-dependent rotor
resistance in the equivalent circuit of Fig. 3.7(b) is the same as
R_r'/s of Fig. 3.7(a) [vide Eqn (3.17)], which also gives the resulting
electromagnetic torque T_e in synchronous watts.

Modified T-form

There is an apparent omission in the circuit model of Fig. 3.3(a),
which comprises impedances only. The circuit models of syn-

chronous machines and dc machines contain a rotational emf term. Even the generalized theory of electrical machines indicates the presence of a rotational emf in the rotor circuit of an induction machine. The conventional circuit model of an induction motor shows the presence of a negative resistance in the rotor circuit, implying generating action at a speed above the synchronous speed. However, the absence of any rotational emf in the rotor branch complicates the explanation of the behaviour of the rotor of an induction machine that serves as a source of power during its generating action, particularly, in the self-excited mode. The concept of a negative resistance seems to offer a computational advantage rather than a convincing explanation.

In the circuit model of Fig. 3.3(a), the stator-referred voltage drop across the slip-dependent resistor in a more complex form is

$$\mathbf{V}_{2,1-s} = \frac{I'_2 R'_r}{s}(1 - s) \tag{3.31}$$

Using Eqn (3.8) to replace $I'_2 R'_r / s$ in Eqn (3.31),

$$\mathbf{V}_{2,1-s} = \mathbf{E}_1(1 - s) + j(s - 1)X'_{lr}\mathbf{I}'_2 \tag{3.32}$$

Referring to Fig. 3.3(a),

$$\mathbf{E}_1 = \mathbf{I}'_2(R'_r + jX'_{lr}) + \mathbf{V}_{2,1-s} \tag{3.33}$$

The combination of Eqns (3.32) and (3.33) yields

$$\mathbf{E}_1 = \mathbf{I}'_2(R'_r + jsX'_{lr}) + \mathbf{E}_1(1 - s) \tag{3.34}$$

The modified circuit model satisfying the rotor-side relation (3.34) is shown in Fig. 3.8.

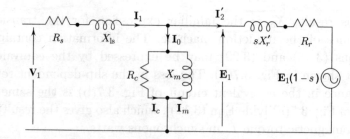

Fig. 3.8 Modified steady-state T-form circuit model

Since

$$E_1 = I_m \omega L_m \tag{3.35}$$

where L_m is the magnetizing inductance, the emf $E_1(1 - s)$ is proportional to the rotor speed and the magnetizing current. This is, therefore, a rotational emf. The total power associated with the rotational emf is

$$
\begin{aligned}
P_e &= 3\text{Re}\left[\mathbf{E}_1(1 - s)\mathbf{I}_2^{'*}\right] \\
&= 3\frac{E_1^2 s(1 - s)R_r'}{R_r'^2 + s^2 X_{lr}'^2}
\end{aligned}
\tag{3.36}
$$

which agrees with Eqn (3.12) in conjunction with Eqn (3.10), and is thus the mechanical power output P_m of the machine ('Re' denotes 'the real part of'). The new circuit model of Fig. 3.8 emphasizes the concept of rotational emf and highlights the similarities between the induction motor, the synchronous motor, and the dc motor.

3.1.6 Effect of Rotor-injected Emf—Slip Power Recovery Scheme

The stator draws a compensating load current I_2' to counteract the rotor mmf in order to sustain the air-gap flux set up by the magnetizing component of the exciting current I_m. Consequently, if the rotor changes its current as a result of any other source of emf in its circuit, the stator would be unable to detect the inclusion of this additional emf in the rotor circuit, as the same change in the rotor current and its power factor can be effected by the inclusion of appropriate values of resistance and inductance (capacitance) in the rotor circuit. Figure 3.9(a) shows the rotor equivalent circuit with an additional emf E_j injected at an angle β with respect to sE_2 and acting in the opposite sense to it. Referring to the phasor diagram in Fig. 3.9(b), it is clear that

$$s\mathbf{E}_2 = \mathbf{I}_2 z_{2s} + \mathbf{E}_j \tag{3.37}$$

Resolving the emfs and the impedance drop along $\mathbf{I}_2$ yields

$$sE_2 \cos \phi_2 = I_2 R_r + E_j \cos(\phi_2 + \beta) \tag{3.38}$$

Multiplying by I_2,

$$sE_2I_2 \cos \phi_2 = I_2^2 R_r + E_j I_2 \cos(\phi_2 + \beta) \qquad (3.39)$$

Referring to the stator,

$$sE_1 I_2' \cos \phi_2 = I_2'^2 R_r' + E_j' I_2' \cos(\phi_2 + \beta) \qquad (3.40)$$

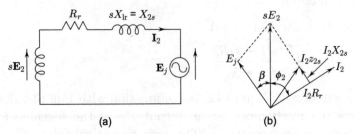

(a) (b)

Fig. 3.9 (a) Rotor circuit model with injected emf; (b) phasor diagram

The right-hand side terms of Eqn (3.40) represent the per-phase electrical power utilized in the rotor circuit. As $E_1 I_2' \cos \phi_2$ represents the air-gap power P_{ag} per phase, it is seen that the slip multiplied by the air-gap power is the sum of the rotor copper losses ($p_{Cu2} = 3I_2'^2 r_2'$) and the power fed into the auxiliary source $[P_2 = 3E_j' I_2' \cos(\phi_2 + \beta)]$. Therefore, the balance of the air-gap power, i.e., $(1 - s)P_{ag}$, must be the mechanical power output P_m.

Adding $(1 - s)E_1 I_2' \cos \phi_2$ to both sides of Eqn (3.40) and considering all the three phases, the electrical power crossing the air-gap can be expressed as

$$P_{ag} = P_{Cu2} + P_2 + P_m \qquad (3.41)$$

where

$$P_m = (1 - s)P_{ag} \qquad (3.42)$$

Multiplying both sides of Eqn (3.40) by $(1 - s)/s$ and substituting in Eqn (3.42) give

$$P_m = 3\frac{1-s}{s}I_2'^2 R_r' + 3\frac{1-s}{s}E_j' I_2' \cos(\phi_2 + \beta) \qquad (3.43)$$

The series of equations (3.38)–(3.43) suggests Fig. 3.10 as the modified version of the per-phase stator-referred conventional equivalent circuit of the induction motor with injected emf in the rotor. Such an induction machine is also known as a *doubly fed induction machine* (DFIM) because of the two power sources employed. Since motoring convention has been followed, P_{ag} and P_m will be negative in the generating mode. P_2 has been considered positive for the power absorbed by the auxiliary source, i.e., for the power flowing out of the slip-ring terminals. The electrical power associated with the slip-dependent secondary resistance and the auxiliary emf, shown in Fig. 3.10, represents the mechanical power.

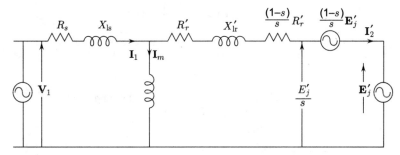

Fig. 3.10 Stator-referred equivalent circuit with injected emf in the rotor

With respect to the flow of power between the motor shaft and the stator, the behaviour of the induction machine, as a result of auxiliary power (P_2) control, can be divided into four operating modes.

Mode I—Subsynchronous motoring operation

In this mode, $s < 1$ and $P_m > 0$. Consequently, according to Eqn (3.42), $P_{ag} > P_m$. $P_{Cu2} + P_2$, which equals sP_{ag}, is positive. For constant-torque operation, P_{ag} is constant, and hence an increase in $P_{Cu2} + P_2$ raises the value of s, which implies a drop in the speed. Figure 3.11(a) shows the power flow diagram in this mode of operation.

Mode II—Supersynchronous motoring operation

In the supersynchronous region, the rotor speed is greater than the synchronous speed, i.e., $s < 0$. For motoring operation, $P_m > 0$, and from Eqn (3.42), P_{ag} is less than P_m and is positive. Consequently, from Eqn (3.41), $P_{Cu2} + P_2$ must be negative. As P_{Cu2} is always positive, P_2 must be negative. Therefore, power must be fed into the slip-ring terminals from the auxiliary source. By increasing the input to the rotor, P_m can be increased. Since P_{ag} remains constant for constant mechanical torque, the machine speed will increase. The power flow diagram for this mode is shown in Fig. 3.11(b).

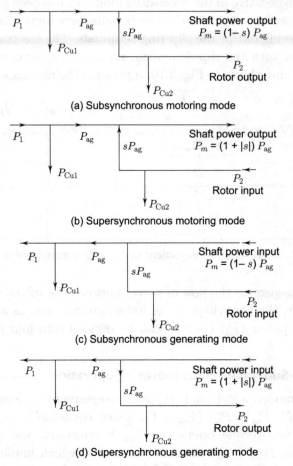

(a) Subsynchronous motoring mode

(b) Supersynchronous motoring mode

(c) Subsynchronous generating mode

(d) Supersynchronous generating mode

Fig. 3.11 Power flow diagram of a doubly fed induction machine

Mode III—Subsynchronous generating action

For generating action, P_m is negative, and as the rotor speed is less than the synchronous speed ($0 < s < 1$), P_{ag} is negative [Eqn (3.42)] and $|P_{ag}| > |P_m|$. Therefore, from Eqn (3.41), $P_{Cu2} + P_2$ must be negative. As P_{Cu2} is positive, P_2 should be made sufficiently negative by injecting power into the rotor circuit in order to make the rotor electrical power sP_{ag} negative. The net electrical power flowing into the grid is, therefore, $P_1 - P_2$. The directions of the various power flows for this generating mode at subsynchronous speeds are shown in Fig. 3.11(c).

Mode IV—Supersynchronous generating action

For supersynchronous operation in the generating mode, the power flow diagram is shown in Fig. 3.11(d). s being negative $(-1 \leq s < 0)$ in this mode, the air-gap power $|P_{ag}|$ is less than the mechanical power $|P_m|$ [Eqn (3.42)], and is negative. The remaining surplus energy sP_{ag} is returned via the rotor circuit to the grid after providing for the secondary losses P_{Cu2}. Mathematically, $sP_{ag} > 0$, both s and P_{ag} being negative.

From the viewpoint of power flow in the rotor circuit, the region of supersynchronous operation in the generating mode can be divided into two subregions as depicted by the torque–speed curves in Fig. 3.5. In subregion I, the rotor speed is less than the rated speed $\omega_{r,\text{rated}}$ but greater than the synchronous speed, and the slip-ring terminals are kept shorted. Rated speed is obtained when the stator carries its rated current. For this configuration, the operating speed range of the generator is small if the stator current is not to exceed its rated value. This is the conventional use of the machine. It is also the most efficient way of operating the machine in this subregion. Any attempt to extract power from the rotor by inserting an external resistance in the rotor circuit will shift the torque–speed curve, and the net output power at a given speed will drop with respect to the conventional use of the same machine. If, however, an electrical source in proper phase is connected to the rotor circuit, the induction machine will be able to feed more power to the supply than with conventional use. This is because the rotor current will then be able to go above the value corresponding to the conventional use of the same machine without exceeding the rated rotor current.

In subregion II, when the rotor speed is higher than the rated speed $\omega_{r,\text{rated}}$, the effective rotor resistance has to be increased to keep the stator current constant and equal to its rated value. The equivalent circuit in Fig. 3.2(c) and Eqns (3.21) and (3.22) suggest that if the stator voltage and frequency are kept constant, then, for constant developed torque and machine current (stator and rotor), the external rotor resistance should be so chosen as to satisfy

$$\frac{R_r}{s_r} = \frac{R_r + R_x}{s} \tag{3.44}$$

Consequently, at any value of slip greater than s_r, the stator power can be held at its rated value by adjusting the rotor resistance in accordance with Eqn (3.44). The additional resistance effect can also be realized by extracting electrical power, equal to the surplus mechanical power, through the slip rings.

3.1.7 Dynamic *d-q* Axis Model

Vector control (i.e., decoupling control) schemes for ac motor drive systems have gained wide acceptance in high-performance, variable-speed applications by creating independent channels for torque and flux controls. In a similar manner, vector control strategies have been proposed for controlling the active and reactive power of the induction generator.

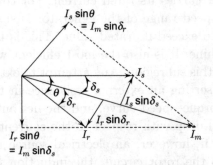

Fig. 3.12 Spatial mmf phasor diagram

Stator and rotor currents flowing through balanced sinusoidally distributed windings set up respective resultant space mmf vectors which may be defined in terms of the current space-vectors I_s and I_r as shown in Fig. 3.12. The developed electromagnetic torque is proportional to the product of the magnitudes of the two current vectors and the sine of their space phase difference, i.e.,

$$\begin{aligned}
T_e &= kI_sI_r \sin\theta \\
&= kI_sI_m \sin\delta_s \\
&= kI_rI_m \sin\delta_r
\end{aligned} \tag{3.45}$$

I_m, the magnetizing current space-vector, represents the resultant air-gap flux vector and $I_s \sin \delta_s$ $(I_r \sin \delta_r)$ represents the torque-producing current vector. These two rotating space-vectors are always in quadrature. The essence of vector control is to force the moving stator and rotor current vectors I_s and I_r to take these magnitudes and positions that enable independent control of I_m and $I_s \sin \delta_s$ $(I_r \sin \delta_r)$. This is achieved by appropriate control of the magnitude and phase of the actual stator (rotor) currents. Vector control makes an induction machine behave like a dc machine with $I_s \sin \delta_s$ $(I_r \sin \delta_r)$ analogous to the armature current and I_m analogous to the field excitation.

The same current vectors I_s and I_r can also be produced by assuming currents flowing through a pair of two orthogonally spaced fictitious identical windings, replacing the original balanced three-phase stator and rotor windings. Such a transformation is known as a *reference frame transformation*. However, for this, mere replacement by a two-phase winding is not sufficient; further insight is necessary in order to develop the mathematical relation between the real and the transformed variables.

Owing to the smooth air gap, the self-inductances of the stator and the rotor windings are constant, but the mutual inductances between them vary with the rotor displacement relative to the stator. This variation of the stator-to-rotor mutual inductances makes the analysis of an induction motor in terms of real variables complicated, as the voltage equations become non-linear. In order to eliminate the effect of the variation of the mutual inductances and, thus, facilitate analysis, a change of variables can be devised for the stator and rotor variables. This gives rise to a fictitious magnetically decoupled two-phase machine, in which the rotor circuits are not only made stationary with respect to the stator circuits but also aligned with the respective stator windings. In this way, all the inductances become constant.

These orthogonally spaced balanced windings, known as the d- and q-windings, may be considered stationary or moving with respect to the stator. Figure 3.13 shows two such sets, one stationary and one rotating. In the stationary reference frame, the d^s and q^s axes are fixed on the stator with either the d^s- or the q^s-axis coinciding with the stator a-phase axis. The rotating d^e-q^e axis may be either fixed on the rotor or made to move at the synchronous speed. If one of the axes of the synchronously rotating reference frame coincides with the air-gap flux vector (i.e, the magnetizing current vector I_m), it is said to be *air-gap flux oriented.*

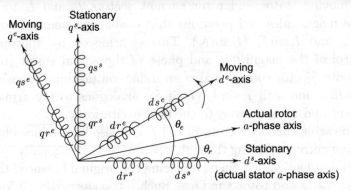

Fig. 3.13 Angular relationships between reference axes

Incidentally, a vector control scheme need not always be designed with respect to the air-gap flux. It may also be designed with respect to the stator or rotor flux with corresponding advantages and limitations. In field-oriented control (FOC), the stator phase currents are first estimated in a synchronously rotating reference frame and then transformed back to the stationary stator frame to feed the machine.

How is the transformation carried out? With the invariance of power as the necessary criterion, and assuming the equivalent two-phase windings to have $\sqrt{3/2}$ times as many turns per phase as the three-phase winding, the fictitious stator d, q, o variables are obtained from the phase variables (a, b, c) through a transform (similar to the Park transform in synchronous machines) defined as

$$
\begin{bmatrix} f^e_{ds} \\ f^e_{qs} \\ f^e_{os} \end{bmatrix} = \sqrt{\frac{2}{3}}
$$

$$
\times \begin{bmatrix} \cos \theta_e & \cos (\theta_e - 2\pi/3) & \cos (\theta_e + 2\pi/3) \\ -\sin \theta_e & -\sin (\theta_e - 2\pi/3) & -\sin (\theta_e + 2\pi/3) \\ 1/\sqrt{2} & 1/\sqrt{2} & 1/\sqrt{2} \end{bmatrix}
$$

$$\times \begin{bmatrix} f_{as} \\ f_{bs} \\ f_{cs} \end{bmatrix} \tag{3.46}$$

where θ_e is the angle of the moving d^e-axis with respect to the stator a-phase winding, as shown in Fig. 3.13.

In Eqn (3.46), f can represent voltage, current, or flux linkage. This transformation is based on the assumption of a distributed sinusoidal winding. The phase variables are obtained from the d, q, o variables through the inverse of the transformation matrix in Eqn (3.46), i.e.,

$$\begin{bmatrix} f_{as} \\ f_{bs} \\ f_{cs} \end{bmatrix} = \sqrt{\frac{2}{3}} \begin{bmatrix} \cos\theta_e & -\sin\theta_e & 1/\sqrt{2} \\ \cos(\theta_e - 2\pi/3) & -\sin(\theta_e - 2\pi/3) & 1/\sqrt{2} \\ \cos(\theta_e + 2\pi/3) & -\sin(\theta_e + 2\pi/3) & 1/\sqrt{2} \end{bmatrix}$$

$$\times \begin{bmatrix} f_{ds}^e \\ f_{qs}^e \\ f_{os}^e \end{bmatrix} \tag{3.47}$$

With reference to Fig. 3.13, replacing θ_e by $\theta_e - \theta_r$ in Eqns (3.46) and (3.47) defines the same transformation for the rotor quantities.

The stator d, q variables in the reference frame fixed to the stator, with the d-axis aligned along the a-phase axis, are related to the phase variables as follows:

$$\begin{bmatrix} f_{ds}^s \\ f_{qs}^s \\ f_{os}^s \end{bmatrix} = \sqrt{\frac{2}{3}} \begin{bmatrix} 1 & -1/2 & -1/2 \\ 0 & \sqrt{3}/2 & -\sqrt{3}/2 \\ 1/\sqrt{2} & 1/\sqrt{2} & 1/\sqrt{2} \end{bmatrix} \begin{bmatrix} f_{as} \\ f_{bs} \\ f_{cs} \end{bmatrix}$$

$$\tag{3.48}$$

In the synchronously rotating reference frame defined by the d^e-q^e axis, the dynamic voltage equations of a three-phase symmetrical induction machine in terms of the equivalent two-phase system, defined by Eqns (3.46) and (3.47), are given by

$$v_{ds}^e = R_s i_{ds}^e + p\lambda_{ds}^e - \omega_e \lambda_{qs}^e \tag{3.49}$$

$$v_{qs}^e = R_s i_{qs}^e + p\lambda_{qs}^e + \omega_e \lambda_{ds}^e \tag{3.50}$$

$$v_{dr}^e = R_r i_{dr}^e + p\lambda_{dr}^e - (\omega_e - \omega_r)\lambda_{qr}^e \tag{3.51}$$

$$v_{qr}^e = R_r i_{qr}^e + p\lambda_{qr}^e + (\omega_e - \omega_r)\lambda_{dr}^e \tag{3.52}$$

The flux linkage equations are

$$\lambda_{ds}^e = L_s i_{ds}^e + M i_{dr}^e$$

$$\lambda_{qs}^e = L_s i_{qs}^e + M i_{qr}^e \tag{3.53}$$

$$\lambda_{dr}^e = L_r i_{dr}^e + M i_{ds}^e$$

$$\lambda_{qr}^e = L_r i_{qr}^e + M i_{qs}^e \tag{3.54}$$

where L_s and L_r are the self-inductances of the stator and the rotor windings, respectively, and M is the mutual inductance between a stator and a rotor winding.

The expression for electromagnetic torque in terms of the currents is

$$T_e = M\frac{P}{2}(i_{qs}^e i_{dr}^e - i_{ds}^e i_{qr}^e) \tag{3.55}$$

Using Eqns (3.53) and (3.54) in Eqn (3.55) gives, respectively, Eqns (3.56) and (3.57), which are sometimes found to be useful.

$$T_e = P/2(\lambda_{ds}^e i_{qs}^e - \lambda_{qs}^e i_{ds}^e) \tag{3.56}$$

$$T_e = P/2(\lambda_{qr}^e i_{dr}^e - \lambda_{dr}^e i_{qr}^e) \tag{3.57}$$

Induction machines are generally operated under balanced conditions. If the terminal voltages form a balanced set, the steady-state currents will also form a balanced set in a symmetrical induction machine. Let the stator terminal voltages be

$$v_a = \sqrt{2}V\cos{(\omega t - \alpha)}$$

$$v_b = \sqrt{2}V\cos\left(\omega t - \alpha - \frac{2\pi}{3}\right)$$

$$v_c = \sqrt{2}V\cos\left(\omega t - \alpha + \frac{2\pi}{3}\right) \tag{3.58}$$

where ω is the angular supply frequency. Similarly,

$$i_a = \sqrt{2}I\cos(\omega t - \beta)$$

$$i_b = \sqrt{2}I \cos(\omega t - \beta - 2\pi/3)$$

$$i_c = \sqrt{2}I \cos(\omega t - \beta + 2\pi/3) \tag{3.59}$$

Substituting Eqns (3.58) and (3.59) into the transformation (3.46) and carrying out the trigonometrical operations on the products $v_{ds}^e i_{ds}^e$, $v_{ds}^e i_{qs}^e$, $v_{qs}^e i_{qs}^e$, and $v_{qs}^e i_{ds}^e$, we get

$$\text{Active power } P = v_{ds}^e i_{ds}^e + v_{qs}^e i_{qs}^e$$

$$= 3VI \cos(\beta - \alpha) \tag{3.60}$$

$$\text{Reactive power } Q = v_{qs}^e i_{ds}^e - v_{ds}^e i_{qs}^e$$

$$= 3VI \sin(\beta - \alpha) \tag{3.61}$$

and

$$(v_{ds}^e)^2 + (v_{qs}^e)^2 = 3V^2 \tag{3.62}$$

For balanced sets, v_o and i_o will be zero. Whether balanced or not, the relations given below always hold good:

$$P = v_{ds}^e i_{ds}^e + v_{qs}^e i_{qs}^e + v_{os}^e i_{os}^e \tag{3.63}$$

$$= v_a i_a + v_b i_b + v_c i_c \tag{3.64}$$

and

$$(v_{ds}^e)^2 + (v_{qs}^e)^2 + (v_{os}^e)^2 = v_{as}^2 + v_{bs}^2 + v_{cs}^2$$

3.2 The Wound-field Synchronous Machine

In wind electric power generation systems, two types of wind turbines are generally used. These are variable-speed and constant-speed turbines. The high-power (750 kW to 2 MW) variable-speed synchronous generator, with field windings on the rotor, is a serious competitor for the wound rotor induction motor. In particular, direct drive variable-speed systems use synchronous machines. As the name indicates, unlike in a wound rotor inducion machine, the rotor of a synchronous machine runs in synchronization with the field produced by the stator winding currents. The salient aspects of the working of a synchronous machine are taken up in the following sections.

3.2.1 Constructional Features

The essential features of a wound-field synchronous machine are presented in the schematic diagram of Fig. 3.14 for a basic two-pole machine with a salient-pole structure. The stator is similar to that of an induction machine. Three identical windings are distributed in slots over the inner surface of the stator with their magnetic axes displaced by 120° from each other. In the figure, the distributed three-phase windings are represented by the three coils aa', bb', and cc'. The rotor is equipped with a field winding df-df' which is fed by a dc source. In a cylindrical rotor, with a smooth air gap, the field winding is distributed, while in a salient-pole rotor with a non-uniform air gap, the field winding comprises concentrated coils. The cylindrical rotor construction, with two or sometimes four poles, is operated at high speed for large-capacity machines, while the salient-pole rotor construction, with a large number of projected poles, is common for slower speed machines. Salient-pole machines are commonly used with wind turbines when the use of a synchronous machine is intended. Sometimes a squirrel cage type winding, called the *amortisseur* or the damper winding, is embedded in the rotor pole face of a salient-pole machine. Assuming sinusoidally distributed mmfs, the damper windings can be represented by two sinusoidally distributed windings, represented by the coils dk-dk' and qk-qk' shown in Fig. 3.14. The

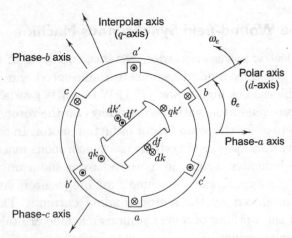

Fig. 3.14 A two-pole, three-phase synchronous machine with salient-pole structure

magnetic axis of the *dk-dk'* winding coincides with the magnetic axis of the main field winding and the magnetic axis of the *qk-qk'* winding is 90° ahead of the *df-df'* winding.

As in the case of an induction machine with *p* pole pairs, balanced three-phase sinusoidal currents flowing through a balanced sinusoidally distributed set of stator windings set up a sinusoidally distributed mmf in the air gap rotating at speed

$$n_s = \frac{f}{p} \; \text{(rps)}$$

Unlike an induction machine, however, the rotor of a synchronous machine rotates at the same angular velocity as the rotating air-gap mmf, and hence the name 'synchronous machine'.

3.2.2 Dynamic Machine Equations

As far as the performance equations are concerned, the round rotor machine can be treated as a special case of the salient-pole machine. In terms of real variables (phase currents and field currents), the analysis of a salient-pole synchronous machine is more complicated compared to that of an induction machine. Not only are the mutual inductances between the stator and the rotor windings functions of the rotor displacement relative to the stator, but also the stator winding self-inductances and the mutual inductances between them are rotor-position-dependent. To make the matter worse, the rotor windings are not identical and the magnetic characteristics along the *d*- and the *q*-axis are different. As a result, the voltage equations are highly non-linear.

To overcome the problems associated with time-varying inductances, the actual stator variables are replaced by a new set of variables, which can be thought of as being associated with three orthogonally spaced fictitious windings ds^e, qs^e, and os^e fixed on the rotor frame as shown in Fig. 3.15. Winding os^e, which is not shown, is in a direction normal to the plane containing the axes of the ds^e and qs^e windings and does not contribute to any mmf in the radial direction in the air gap. As the fictitious and rotor windings are not in relative motion, the mutual inductances between them are all constants. Also the *d*-axis windings are magnetically decoupled from the *q*-axis windings.

The matrix equation of the transformation from the actual three-phase stator variables to the new stator variables in the rotor

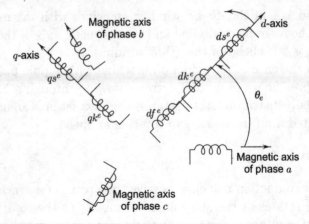

Fig. 3.15 Rotating d-q axes windings and stationary phase windings

reference frame (rotating d-q-o axes) is given by Eqn (3.46). The equations of the inverse transformation are given by Eqn (3.47). Being on the rotor reference frame, variables associated with the rotor windings need no transformation.

In the rotor reference frame, the voltage equations for the field windings and the substituted stator windings may be written as

$$v_{ds}^e = R_s i_{ds}^e + p\lambda_{ds}^e - \lambda_{qs}^e \omega_e$$

$$v_{qs}^e = R_s i_{qs}^e + p\lambda_{qs}^e + \lambda_{ds}^e \omega_e$$

$$v_{os}^e = R_s i_{os}^e + p\lambda_{os}^e$$

$$0 = R_k' {i_{dk}'}^e + p\lambda_{dk}^{e\prime} \qquad (3.65)$$

$$0 = R_k' {i_{qk}'}^e + p\lambda_{qk}^{e\prime}$$

and

$$v_f' = R_f' i_f' + p\lambda_f'$$

where

$$\lambda_{ds}^e = L_d i_{ds}^e + L_{md}({i_f'}^e + i_{dk}'^e)$$

$$\lambda_{qs}^e = L_q i_{qs}^e + L_{mq} i_{qk}'^e$$

$$\lambda_{os}^e = L_{ls} i_{os}^e$$

$$\lambda_{dk}'^{e} = L_{md}(i_{ds}^{e} + i_{f}'^{e}) + L_{dk}'i_{dk}'^{e} \tag{3.66}$$

$$\lambda_{qk}'^{e} = L_{mq}i_{qs}^{e} + L_{qk}'i_{qk}'^{e}$$

and $\quad \lambda_{f}'^{e} = L_{f}'i_{f}'^{e} + L_{md}(i_{ds}^{e} + i_{dk}'^{e})$

In these equations, the rotor winding quantities are referred to the stator. Substitution of the flux linkage expressions (3.66) into Eqn (3.65) yields the voltage equations in terms of the currents as follows:

$$
\begin{bmatrix} v_{ds}^{e} \\ v_{qs}^{e} \\ v_{os}^{e} \\ 0 \\ 0 \\ v_{f}' \end{bmatrix}
=
\begin{bmatrix}
R_s + pL_d & -\omega_e L_q & 0 & pL_{md} & -\omega_e L_{mq} & pL_{md} \\
\omega_e L_d & R_s + pL_q & 0 & \omega_e L_{md} & pL_{mq} & \omega_e L_{md} \\
0 & 0 & R_s + pL_{ls} & 0 & 0 & 0 \\
pL_{md} & 0 & 0 & R_{dk}' + pL_{dk}' & 0 & pL_{md} \\
0 & pL_{mq} & 0 & 0 & R_{qk}' + pL_{qk}' & 0 \\
pL_{md} & 0 & 0 & pL_{md} & 0 & R_f' + pL_f'
\end{bmatrix}
\times
\begin{bmatrix} i_{ds}^{e} \\ i_{qs}^{e} \\ i_{os}^{e} \\ i_{dk}'^{e} \\ i_{qk}'^{e} \\ i_{f}'^{e} \end{bmatrix}
\tag{3.67}
$$

The expression for the electromagnetic torque in d^{e}-q^{e} variables may be obtained from the emf equations (3.65) by multiplying the rotational emf terms $-\omega_e\lambda_{qs}^{e}$ and $\omega_e\lambda_{ds}^{e}$ by i_{ds}^{e} and i_{qs}^{e}, respectively, adding them, and then dividing the sum by the mechanical speed ω_r of the rotor, i.e.,

$$T_e = \frac{P}{2}(\lambda_{ds}^{e}i_{qs}^{e} - \lambda_{qs}^{e}i_{ds}^{e}) \tag{3.68}$$

which, according to Eqn (3.66), may be written as

$$T_e = \frac{P}{2}\left[L_{md}(i_{ds}^e + i_f'^e + i_{dk}'^e)i_{qs}^e - L_{mq}(i_{qs}^e + i_{qk}'^e)i_{ds}^e\right]$$

(3.69)

The electromagnetic torque is positive for motor action, as motoring convention has been followed in the emf equations. The torque and the speed are related as follows:

$$T_e = \frac{2J}{P}\frac{d\omega_e}{dt} + T_l$$

(3.70)

T_l is positive for torque load under motor action and negative for torque input under generator action.

3.2.3 Steady-state Operation

Steady-state d-q axis equations

Let us consider the no-load operation of the machine from a symmetrical three-phase system of sinusoidal voltages of angular frequency ω. Counting time from the instant the phase-a voltage is positive and maximum, the stator voltages will be

$$v_{as} = V_m \cos\omega t$$

$$v_{bs} = V_m \cos(\omega t - 120°)$$

(3.71)

$$v_{cs} = V_m \cos(\omega t + 120°)$$

This balanced set may be regarded to have set up a voltage space-vector rotating at a constant angular velocity ω, making an angular displacement ωt with respect to the chosen time-zero position, i.e., the a-phase axis, as shown in Fig. 3.16(a). Substituting Eqns (3.71) into the equations of transformation [Eqn (3.46)] yields

$$v_{ds}^e = \sqrt{\frac{3}{2}}V_m \cos(\omega t - \theta_e)$$

and

$$v_{qs}^e = \sqrt{\frac{3}{2}}V_m \sin(\omega t - \theta_e)$$

(3.72)

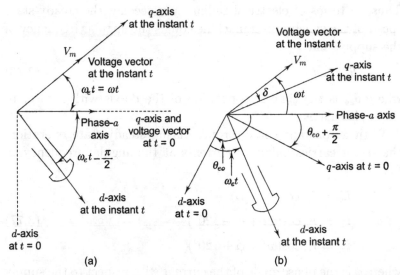

Fig. 3.16 Positions of the d-q axes: (a) with no load and (b) with load

From Eqn (3.67), for no-load ($i_d = i_q = 0$) steady-state operation ($i'_{dk} = i'_{qk} = 0$) we have

$$v_{ds}^e = 0$$

and

$$v_{qs}^e = L_{md}\omega_e i'_f \tag{3.73}$$

A comparison of Eqns (3.72) and (3.73) suggests that

$$\theta_e = \omega t - \frac{\pi}{2} \tag{3.74}$$

and

$$V_m = \sqrt{\frac{2}{3}} L_{md}\omega_e i'_f \tag{3.75}$$

Equation (3.75) gives the maximum value of the excitation emf E_{fm} per phase for the field current i'_f. By definition (Fig. 3.14), θ_e is the position of the d-axis with reference to the a-phase axis. Therefore, under the no-load condition, the q-axis coincides with the voltage space-vector, as shown in Fig. 3.16(a).

It follows from Eqn (3.74) that

$$\frac{d\theta_e}{dt}(= \omega_e) = \omega$$

Thus, in terms of electrical radians per second, the steady-state speed of the synchronous machine equals the angular frequency of the supply, and

$$\theta_e = \omega t + \theta_{eo} \tag{3.76}$$

where θ_{eo} is the time-zero position of the d-axis with respect to the a-phase axis.

With a constant shaft torque, during steady-state operation, the stator carries balanced currents at the angular frequency ω. Let

$$i_{as} = I_m \cos(\omega t + \phi)$$
$$i_{bs} = I_m \cos(\omega t + \phi - 120°) \tag{3.77}$$
$$i_{cs} = I_m \cos(\omega t + \phi + 120°)$$

where ϕ is the phase angle of the current with respect to the supply voltage.

Substitution of Eqns (3.77) in the transformation equations (3.46) and subsequent reduction yield

$$i_{ds}^e = \sqrt{\frac{3}{2}} I_m \cos(\omega t + \phi - \theta_e)$$
$$i_{qs}^e = \sqrt{\frac{3}{2}} I_m \sin(\omega t + \phi - \theta_e) \tag{3.78}$$
$$i_{os}^e = 0$$

Therefore, from Eqns (3.78) and (3.76),

$$i_{ds}^e = \sqrt{\frac{3}{2}} I_m \cos(\phi - \theta_{eo})$$
$$i_{qs}^e = \sqrt{\frac{3}{2}} I_m \sin(\phi - \theta_{eo}) \tag{3.79}$$

It is clear from Eqns (3.79) that i_{ds} and i_{qs} are dc values; these are henceforth denoted by capital letters.

It follows from Eqns (3.79) that the rms value of the stator current is given by

$$I_a = \frac{\sqrt{I_{ds}^{e2} + I_{qs}^{e2}}}{\sqrt{3}} \tag{3.80}$$

Substituting Eqns (3.79) in Eqns (3.65), we find that

$$V_{ds}^e = R_s I_{ds}^e - \omega_e \lambda_{qs}^e$$

and

$$V_{qs}^e = R_s I_{qs}^e + \omega_e \lambda_{ds}^e \tag{3.81}$$

From Eqns (3.72) and (3.76), V_{ds}^e and V_{qs}^e can also be expressed as

$$V_{ds}^e = \sqrt{\frac{3}{2}} V_m \cos(\theta_{eo})$$

and

$$V_{qs}^e = -\sqrt{\frac{3}{2}} V_m \sin(\theta_{eo}) \tag{3.82}$$

It is evident from Eqns (3.81) that both V_{ds} and V_{qs} exist for a non-zero shaft torque. R_s being small, for motor operation, V_{ds} is negative and V_{qs} is definitely positive. Thus it is apparent from Eqns (3.72) that

$$\omega t - \theta_e > \frac{\pi}{2}$$

that is,

$$\theta_e < \omega t - \frac{\pi}{2} \tag{3.83}$$

Thus, when compared with Eqn (3.74) and the axis alignments in Fig. 3.16(a), the condition in Eqn (3.83) reveals that the d-q axes under the shaft torque must fall behind their no-load position for the same voltage vector position with respect to the phase-a axis. This is shown in Fig. 3.16(b). The displacement of the q-axis relative to the voltage space-vector is known as the *rotor angle*, or the *torque angle*, δ.

Torque angle

The concept of the torque angle introduced above holds good under all conditions, dynamic or steady state, under generator or motor operation. With reference to Fig. 3.16(b), the torque angle δ in the general sense is given as

$$\delta = \int_0^t \left[\omega_e(\xi) - \omega(\xi)\right] d\xi + \theta_{eo} + \frac{\pi}{2} \tag{3.84}$$

Under steady-state operation the rotor speed equals the synchronous speed. Equation (3.70) implies that the load torque is balanced by the electromagnetic torque T_e. If now the load torque T_l is increased in the positive direction, a torque imbalance occurs. It can be seen from the torque balance equation (3.70) that the rotor decelerates, making $\omega_e < \omega$, whereupon $|\delta|$ increases in accordance with Eqn (3.84). T_e changes in the positive direction until it equals T_l at some increased negative value of δ. Equilibrium is not yet reached. Though the decelerating torque at this value of δ is zero, the rotor is still running at a speed below the synchronous speed. The angle $|\delta|$ continues to increase, forcing T_e to increase. As a consequence of Eqn (3.70), acceleration starts and the speed of the rotor increases to the synchronous speed when T_e is at its highest value. Now the rotor accelerates above the synchronous speed, and the magnitude of the rotor angle starts decreasing as the integrand in Eqn (3.84) becomes positive. The ensuing process involves oscillation about the synchronous speed with an ultimate increase in the negative value of the torque angle δ matching the developed electromagnetic torque with the shaft load torque at the steady synchronous speed.

Under the steady-state condition, when the time-zero position is selected for v_{as} to be positive maximum at $t = 0$, the torque angle

$$\delta = \theta_{eo} + \frac{\pi}{2} \tag{3.85}$$

Steady-state torque equation

From Eqn (3.69) the electromagnetic torque under the steady-state condition is

$$T_e = \frac{P}{2}\left[(L_{md} - L_{mq})I_{ds}^e I_{qs}^e + L_{md}I_{qs}^e I_f^{'e}\right] \tag{3.86}$$

Combining Eqns (3.81) and (3.82) and using Eqn (3.85), we have

$$R_s I_{ds}^e - \omega_e L_q I_{qs}^e = \sqrt{\frac{3}{2}}V_m \sin\delta \tag{3.87}$$

$$R_s I_{qs}^e - \omega_e(L_d I_{ds}^e + L_{md}I_f^{'e}) = \sqrt{\frac{3}{2}}V_m \cos\delta \tag{3.88}$$

If Eqns (3.87) and (3.88) are solved for I_{ds} and I_{qs}, ignoring R_s, and the results are used in Eqn (3.86) along with Eqn (3.75), T_e can be expressed as

$$T_e = -\frac{3}{2}\frac{P}{\omega_e}\left[\frac{E_f V}{X_d}\sin\delta + \frac{1}{2}\left(\frac{1}{X_q} - \frac{1}{X_d}\right)\sin 2\delta\right] \qquad (3.89)$$

where E_f and V are the per-phase rms values $E_{fm}/\sqrt{2}$ and $V_m/\sqrt{2}$ of the excitation emf and the terminal voltage, respectively.

Steady-state circuit model of a non-salient-pole synchronous machine

The equivalent circuit model is often found to be very convenient to study, with sufficient reliability, the performance characteristics of a machine, particularly under steady-state operation. Under steady state, whatever the load may be, the damper windings carry no current and the field current undergoes no change. Hence their impedances need not be considered in the steady-state equivalent circuit. Let us consider the generating operation.

A field current I_f sets up the flux ϕ_f which, owing to the rotation of the rotor, induces an emf E_f in the stator winding. When the stator winding is closed, the emf E_f in the stator winding drives a current I_a through the stator winding. I_a sets up the flux ϕ_a, which in turn induces an emf E_a in the stator winding lagging behind I_a by 90°. From the phasor diagram in Fig. 3.17(a), the net terminal voltage is

$$\mathbf{V}_t = \mathbf{E}_f + \mathbf{E}_a$$

i.e.,

$$\mathbf{E}_f = \mathbf{V}_t - \mathbf{E}_a \qquad (3.90)$$

As E_a is proportional to I_a and $-\mathbf{E}_a$ leads $\mathbf{I}_a$ by 90°, $-\mathbf{E}_a$ in Eqn (3.90) may be replaced by a drop across a reactance X_s, as shown in Fig. 3.17(c). Here X_s is called the *synchronous reactance*.

A significant part ϕ_{ar} of ϕ_a crosses the air gap and links the field winding. The remaining part ϕ_{al} links the stator winding only. ϕ_{ar} and ϕ_{al} are known, respectively, as the armature reaction flux and the armature leakage flux. Each component of the flux induces an emf of its own, i.e., E_{ar} by ϕ_{ar} and E_{al} by ϕ_{al}. Hence, the synchronous reactance X_s can be divided into two components, X_{ar} and X_{al}, the drops across which are the measure of the emfs

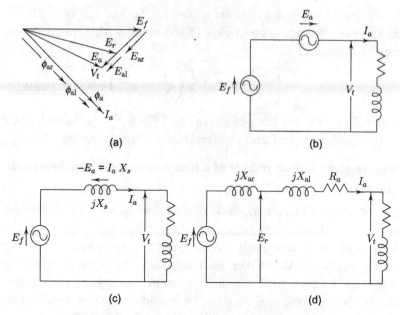

Fig. 3.17 Development of the equivalent circuit

E_{ar} and E_{al} induced, respectively, by the component fluxes ϕ_{ar} and ϕ_{al}. Now ϕ_f and ϕ_{ar} combine to give the resultant flux ϕ_r, which induces E_r, the *air-gap voltage*, in the stator winding. Hence, in the equivalent circuit, E_r is obtained by subtracting jI_aX_{ar} from $\mathbf{E}_f$, as shown in Fig. 3.17(d). Finally, when the armature resistance is included, the per-phase equivalent circuit is represented by Fig. 3.17(d).

3.2.4 Steady-state Model with Rectifier Load

The synchronous generator in a variable-speed wind energy conversion system is connected to a dc link through a controlled, or uncontrolled, bridge rectifier. From this dc link, containing a reactor or a capacitor, power is fed into the utility system by a bridge inverter. The dc link decouples the variable-frequency generator from the constant-frequency grid system. The schematic diagram of a circuit configuration of the generator loaded with a rectifier can be drawn as shown in Fig. 3.18. V_I is the dc-link voltage at the inverter input side. The d, q variables of the machine in steady state are constants. Therefore, setting $p = 0$

in Eqn (3.67), the steady-state armature voltage equations in the rotor reference frame for the generating operation can be written as

$$V_{ds}^e = -R_s I_{ds}^e + \omega_e L_q I_{qs}^e \tag{3.91}$$

and

$$V_{qs}^e = -R_s I_{qs}^e + \omega_e (L_{md} I_f' - L_d I_{ds}^e) \tag{3.92}$$

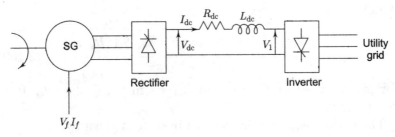

Fig. 3.18 Schematic diagram of the synchronous generator fed ac/dc/ac link

The use of Eqn (3.85) in Eqns (3.79) and (3.82) gives the following equations relating the steady-state d, q and the three-phase stator variables in terms of the phase and torque angles:

$$V_{ds}^e = \sqrt{3}V \sin \delta, \qquad V_{qs}^e = \sqrt{3}V \cos \delta \tag{3.93}$$

and

$$I_{ds}^e = -\sqrt{3}I \sin(\phi - \delta), \qquad I_{qs}^e = \sqrt{3}I \cos(\phi - \delta) \tag{3.94}$$

The condition of power balance yields

$$V_{ds}^e I_{ds}^e + V_{qs}^e I_{qs}^e = R_{dc} I_{dc}^2 + V_I I_{dc} \tag{3.95}$$

Substituting Eqns (3.91) and (3.92) in Eqn (3.95) to eliminate V_{ds}^e and V_{qs}^e, and using Eqn (3.69),

$$R_{dc} I_{dc}^2 + V_I I_{dc} = -R_s (I_{ds}^{e2} + I_{qs}^{e2}) - 2T_e \frac{\omega_e}{P} \tag{3.96}$$

On computation, T_e will assume a negative value for the generating operation. Squaring and adding the equations in Eqn (3.94) give

$$I_{ds}^{e2} + I_{qs}^{e2} = 3I^2 \tag{3.97}$$

For level dc-link current I_{dc}, the rms value of the fundamental component of the generator current is

$$I = \frac{\sqrt{6}}{\pi} I_{dc} \tag{3.98}$$

The use of Eqns (3.97) and (3.98) in Eqn (3.96) yields

$$\left(R_{dc} + \frac{18}{\pi^2} R_s\right) I_{dc}^2 + V_I I_{dc} + \frac{2T_e}{P} w_e = 0 \tag{3.99}$$

For the condition $I_{dc} \geq 0$,

$$I_{dc} = \frac{-V_I + \sqrt{V_I^2 - (8T_e w_e/P)(R_{dc} + 18R_s/\pi^2)}}{2(R_{dc} + 18/\pi^2)} \tag{3.100}$$

For given values of T_e, V_I, and w_e, we may calculate I_{dc} from Eqn (3.100).

The dc-link equation in the steady state is given by

$$V_{dc} = R_{dc} I_{dc} + V_I \tag{3.101}$$

For the controlled rectifier,

$$V_{dc} = \frac{3\sqrt{6}}{\pi} V \cos\alpha - \frac{3}{\pi} w_e L_d'' I_{dc} \tag{3.102}$$

and

$$V = \frac{\pi}{3\sqrt{6}} \frac{V_{dc}}{\cos\alpha} + \frac{w_e L_d'' I_{dc}}{\sqrt{6} \cos\alpha} \tag{3.103}$$

where L_d'' is the sub-transient reactance of the machine. The fundamental power factor of the generator is

$$\cos\phi = \frac{1}{2} [\cos\alpha + \cos(\alpha + \mu)] \tag{3.104}$$

The overlap angle μ for a six-pulse rectifier may be worked out from the relation

$$w_e L_d'' I_{dc} = \sqrt{\frac{3}{2}} V [\cos\alpha - \cos(\alpha + \mu)]$$

Once I_{dc} is calculated from Eqn (3.100), I, V_{dc}, and V are obtained from Eqns (3.98), (3.101), and (3.103), respectively. Substitution of Eqns (3.93) and (3.94) in Eqn (3.91) and the consequent rearrangement of the terms yield

$$\tan \delta = \frac{IR_s \sin \phi + I\omega_e L_q \cos \phi}{V + IR_s \cos \phi - I\omega_e L_q \sin \phi} \qquad (3.105)$$

Using the calculated values of V, I, and $\cos \phi$ in Eqn (3.105), the torque angle δ is obtained. Equations (3.93) and (3.94) can now be used to get the values of I_{ds}^e, I_{qs}^e, V_{ds}^e, and V_{qs}^e. I_f' is then determined from Eqn (3.92).

3.3 The Permanent Magnet Synchronous Machine

3.3.1 Constructional Aspects

Wind turbines run at inconveniently low speeds, typically 25–50 rpm. A speed-increasing gear box is required to run induction machines and conventional synchronous machines at 1000 or 1500 rpm for operation with the utility network. Additional cost, weight, power loss, regular maintenance, and noise generation are some of the problems associated with the gear box. This speed boost is necessary, as induction and synchronous machines cannot be built with pole pitches less than 150 mm, and a large number of poles in the range 120–240, necessary for the direct-coupled generator turning at low speed, cannot be accommodated within an acceptable diameter of the generator, which should fit inside the nacelle with the gear box. Therefore, low-speed, direct-coupled generators are required, particularly for turbines with a large diameters.

Permanent magnet (PM) excitation considerably brings down the pole pitch requirement, which should be less than 40 mm. This allows the rotor to be within an acceptable diameter, which makes the housing of the generator inside the nacelle possible. Several rotor configurations of permanent magnet machines have been developed. Some typical ones are illustrated here. In the surface-type permanent magnet machine, high-energy, rare-earth magnets such as neodymium-iron-boron (Nd-Fe-B) are mounted on the rotor surface, as shown in Fig. 3.19. In an interior-type machine, as shown in Fig. 3.20, cheaper ferrite magnets are circumferentially oriented between flux-concentrating pole pieces. The surface-type machine has lower structural integrity and mechanical robustness. The former type gives equal *d*- and *q*-axis reactances while the latter has a somewhat greater *q*-axis reactance than the *d*-axis reactance. In per-unit terms, both the reactance values are small because of the large number of poles. This provides the PM machines high peak torque capability to resist higher-than-rated torque for short periods during wind gusts and repeated torque pulsations of up to 20% of the rated torque.

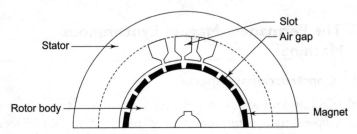

Fig. 3.19 Cross section of a surface-mounted PM machine

Figure 3.21 shows the machine topology of an axial flux machine. Two rotor discs carry magnets in an N-S-N-S-... order with like poles on the discs facing each other. A circumferentially wound iron tape forms the slotless stator with coils surrounding the core. Other topologies of axial flux machines also exist.

The flux distribution from the permanent magnet pole is approximately rectangular, and windings with low harmonic winding factors are necessary for a near sinusoidal emf. The pole pitch

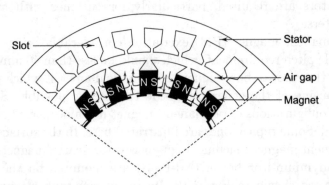

Fig. 3.20 Cross section of the interior-type (buried) PM machine (circumferential magnet motor)

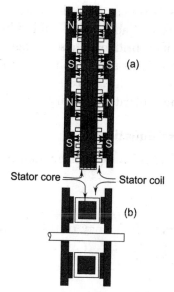

Fig. 3.21 Axial flux machine: (a) unfolded view, (b) schematic diagram of the cross section (torus topology)

being small and the practical value of the slot pitch being of the order of 12–15 mm, a conventional winding with three slots per pole per phase is not possible. The fractional slot winding, which gives favourable fundamental and harmonic winding factors, is used.

In one type of construction, shown in Fig. 3.22, the stator core comprises a large number of separate E-shaped cores arranged circumferentially. Each core carries a coil on the central limb.

Fig. 3.22 A radial flux permanent magnet machine with E-shaped cores making up the stator

Mutual inductances between the coils being very small, the coils operate independently. Each coil is individually connected to a single-phase rectifier bridge. These bridges are connected in parallel to the dc link. This avoids complex interconnection of the coils for three-phase windings. Rotors are made of ferrite magnet blocks with opposing polarity to form the multi-pole rotor.

3.3.2 Steady-state Equations

The generated emf E_g of a permanent magnet generator can be expressed as

$$E_g = k_E \omega \tag{3.106}$$

where ω is the angular frequency of the generator and is related to the mechanical speed as $\omega = P\omega_r/2$.

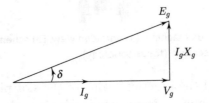

Fig. 3.23   The phasor diagram

Assuming unit power factor and referring to the phasor diagram shown in Fig. 3.23, the relationship between the stator terminal voltage V_g and the current I_g is

$$\mathbf{E}_g = \mathbf{V}_g + j\mathbf{I}_g X_s \tag{3.107}$$

The generated power can be expressed as

$$P_g = 3V_g I_g = 3\frac{V_g E_g}{X_s}\sin\delta \tag{3.108}$$

From the phasor diagram in Fig 3.23,

$$V_g = E_g \cos\delta$$

i.e.,

$$\delta = \arccos\frac{V_g}{k_E\omega} \tag{3.109}$$

Using Eqn (3.109) in Eqn (3.108), the power equation can be expressed as

$$P_g = \frac{3}{2}\frac{E_g^2}{X_s}\sin 2\delta \tag{3.110}$$

Using Eqn (3.106) in Eqn (3.110), the equation representing the torque can be expressed as

$$T_g = T_{\max}\sin 2\delta \tag{3.111}$$

where

$$T_{\max} = \frac{3}{4}\frac{k_E^2 P}{L_s} \tag{3.112}$$

3.4 Power Flow Between Two Synchronous Sources

A pulse width modulated (PWM) voltage source inverter is used to exchange power between a variable-frequency source and a fixed-frequency ac system through a dc link. The inverter produces an output voltage V_I at the fundamental frequency with the required phase angle and magnitude and synchronized with the ac system voltage V_s through an inductor. Under the assumption of balanced sinusoidal voltages, the per-phase steady-state equivalent circuit for the synchronized inverter–ac system and its phasor diagram can be drawn as shown in Fig. 3.24. From the phasor diagram [see Fig. 3.24(b)], the following relations are obtained. Power flow into the system,

$$P_s = 3V_s I_s \cos\phi$$

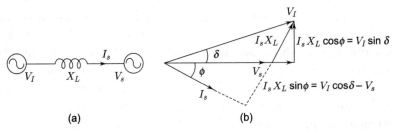

(a) (b)

Fig. 3.24 (a) Schematic diagram of an inverter/ac system interface; (b) phasor diagram

$$= 3\frac{V_s}{X_L} I_s X_L \cos\phi$$

$$= 3\frac{V_s}{X_L} V_I \sin\delta \qquad (3.113)$$

Reactive power flow,

$$Q_s = 3V_s I_s \sin\phi$$

$$= 3\frac{V_s}{X_L} I_s X_L \sin\phi$$

$$= 3\frac{V_s}{X_L} V_I \cos\delta - 3\frac{V_s^2}{X_L} \qquad (3.114)$$

Thus, the real power and the reactive power flow can be changed by controlling the inverter output voltage and its phase angle relative to the ac system.

3.5   Induction Generator Versus Synchronous Generator

Synchronous generators and induction generators have their own merits and limitations. The balance between the advantages and disadvantages of the two is situation-specific. At first glance, a cage induction machine is preferred for its ruggedness and low cost compared to a synchronous generator.

The induction machine, coupled to the utility system, finds favour for fixed-pitch, nearly constant speed wind turbines in order to provide damping for the wind turbine drive train when it faces fluctuation in the power input due to wind speed variations. An induction generator has no synchronization problem. It has relaxed stability criteria and is practically free from hunting owing to the presence of damping, given by the slope of the induction machine torque–slip curve. On the other hand, transient stability may be a serious problem with the grid-connected synchronous machine as it represents a stiff compliance in the dynamic model of the wind turbine drive train. Damping must be provided for either in the machine or mechanically in the drive train.

For symmetrical faults, an induction generator does not contribute any fault current to the network except instantaneous fault current, while for unbalanced faults the contribution is sustained.

Be it a balanced or an unbalanced fault, a synchronous generator contributes fault current.

An induction generator demands lagging reactive VA of the order of 30% of its output kVA from the utility system and produces a current surge at the time of switching-in and during acceleration, both these adversely affecting the voltage of a network with a low fault level. Against this background, a synchronous generator has a low excitation demand, and its active and reactive power can be adjusted by pitch control and field control. In autonomous systems, voltage adjustment by field control is simple.

Worked Examples

1. A 75-kW, three-phase, 400-V, 50-Hz, four-pole wound rotor induction motor, with star-connected stator windings, runs at a slip of 3.5% under rated shaft torque. It has the following per-phase equivalent-circuit parameters referred to the stator: $R_s = 0.075\ \Omega$, $R'_r = 0.06\ \Omega$, $X_{ls} = 0.15\ \Omega$, $X'_{lr} = 0.12\ \Omega$, $X_m = 6.1\ \Omega$. A constant rotational loss of 1.7 kW may be assumed for operation at speeds close to the synchronous speed.

The induction machine is now driven at a speed of 1530 rpm (−2% slip) with the stator terminals remaining connected to the 400-V supply. Find the power fed into the supply for each of the following conditions.

(a) Rotor terminals short-circuited,
(b) 1.15 V per phase (referred to the stator) injected into the rotor, and in phase with the supply, and
(c) 1.15 V per phase (referred to the stator) injected into the rotor and phase shifted by 180° with respect to the supply.

Solution

We solve the problem with reference to the equivalent circuit shown in Fig. 3.10. Application of Thevenin's theorem to the stator part of the circuit permits the replacement of the equivalent circuit by the circuit shown in Fig. 3.25.

In the circuit of Fig. 3.25, the equivalent source voltage is

$$\mathbf{V}_{\text{Th}} = \frac{\mathbf{V}_1 \angle 0° \times jX_m}{R_s + j(X_{ls} + X_m)} = V_{\text{Th}} \angle \theta_{\text{Th}}$$

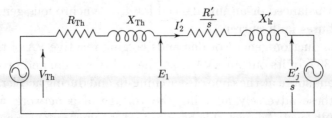

Fig. 3.25   Induction motor equivalent circuit using Thevenin's theorem

and the equivalent stator impedance is

$$Z_{\text{Th}} = \frac{(R_s + jX_{\text{ls}})(jX_m)}{R_s + j(X_{\text{ls}} + X_m)}$$

Simultaneous reference to the circuits in Figs 3.10 and 3.25 furnishes the following equations. Stator-referred rotor current,

$$\mathbf{I}_2' = \frac{V_{\text{Th}} \angle \theta_{\text{Th}} - \frac{E_j'}{s} \angle \theta_j}{(R_{\text{Th}} + R_r'/s) + j(X_{\text{Th}} + X_{\text{lr}}')} = I_2' \angle \phi_2 \qquad (3.115)$$

Stator induced emf,

$$\mathbf{E}_1 = \mathbf{I}_2' \left(\frac{R_r'}{s} + jX_{\text{lr}}' \right) + \frac{\mathbf{E}_j'}{s} = E_1 \angle \theta_m \qquad (3.116)$$

Magnetizing current,

$$\mathbf{I}_m = \frac{\mathbf{E}_1}{jX_m} \qquad (3.117)$$

Stator current,

$$\mathbf{I}_1 = \mathbf{I}_2' + \mathbf{I}_m = I_1 \angle \phi_1 \qquad (3.118)$$

Stator input power,

$$P_1 = 3V_1 I_1 \cos \phi_1 \qquad (3.119)$$

Electrical power output through the slip rings,

$$P_2 = 3E_j' I_2' \cos(\theta_j - \phi_2) \qquad (3.120)$$

Developed mechanical power,

$$P_m = 3\frac{1-s}{s} I_2'^2 R_r' + 3\frac{1-s}{s} E_j' I_2' \cos(\theta_j - \phi_2) \qquad (3.121)$$

Net power from the supply,

$$P_o = P_1 - P_2 \tag{3.122}$$

Shaft power,

$$P_{\text{shaft}} = P_m - P_{m,\text{loss}} \tag{3.123}$$

Efficiency,

$$\eta_n = \frac{P_{\text{shaft}}}{P_o} \quad \text{for the motoring mode} \tag{3.124}$$

$$\eta_G = \frac{P_o}{P_{\text{shaft}}} \quad \text{for the generating mode} \tag{3.125}$$

Substitution of the numerical values of the parameters and sequential use of Eqns (3.115)–(3.125) yield the following rated currents for the stator and the rotor when the machine, with short-circuited slip rings (i.e., $E_j = 0$), is operated as a motor at its rated voltage ($V_1 = 230\angle 0°, 50$ Hz) and speed ($s = 0.035$):

$$I_1 = 134.8 \text{ A}, \quad I_2' = 127.7 \text{ A}$$

Proceeding in the same manner, the stator and stator-referred rotor currents, and the stator and shaft power, under generating mode, for $s = -0.033$, are obtained as follows:

$$I_1 = 135.0 \text{ A}, \quad I_2' = 128.0 \text{ A}$$

$$P_1 = -83.9 \text{ kW}, \quad P_{\text{shaft}} = -93.5 \text{ kW}$$

Hence, under the generating mode, the rated slip s_r shown in Fig. 3.5 is -0.033. Using Eqns (3.115)–(3.125) sequentially, the following results are obtained for the operation of the induction machine in the generating mode under the stipulated conditions, the slip in each case being $s = -0.02$, corresponding to the supersynchronous speed 1530 rpm:

$E_j' \angle \theta_j$ (V)	I_1 (A)	I_2' (A)	P_1 (kW)	P_2 (kW)	P_o (kW)	P_{shaft} (kW)	η_G (p.u.)
0	86.4	76.3	−50.75	0.0	−50.75	−56.0	0.91
$1.15 + j0.0$	103.9	95.9	−63.85	−0.33	−63.52	−70.1	0.91
$-1.15 + j0.0$	69.7	56.8	−37.65	0.20	−37.85	−42.0	0.90

The results given in this table demonstrate certain important points. When the slip is less than its rated value, more power can be extracted from the stator, without exceeding the rated currents, by injecting power into the rotor circuit. On the other hand, if power is extracted through the slip rings, the net power fed into the grid decreases at the same value of the slip. However, if the slip goes beyond the rated value, power has to be extracted through the rotor to keep the machine currents below their rated values.

2. In the per-phase equivalent circuit model of a star-connected, 415-V, 50-Hz, six-pole squirrel cage induction motor, the impedances are as follows: $R_s = 0.085\ \Omega$, $R'_r = 0.07\ \Omega$, $X_m = 7\ \Omega$, $X_{ls} = 0.18\ \Omega$, $X'_{lr} = 0.18\ \Omega$. The machine runs at a slip of -0.025, supplying power to a utility network at 415 V, 50 Hz.

(a) Find the values of the stator and the rotor currents, the electromagnetic torque, and the power input using a synchronously rotating stator voltage vector oriented d-q axis model. Also, draw the space-phasor diagram showing the relative positions of the voltage, current, and flux linkage vectors of the stator and the rotor.

(b) Obtain the d- and q-axis components of the voltage current and flux vectors for orientation in the stator flux vector direction.

Solution

(a) The peak value of the supply phase voltage and the machine inductances are

$$V_m = \sqrt{\frac{2}{3}} 415 = 338.8\ \mathrm{V}$$

$$L_s = L'_r = \frac{7 + 0.18}{2\pi \times 50} = 22.85\ \mathrm{mH}$$

and

$$M = \frac{7}{2\pi \times 50} = 22.3\ \mathrm{mH}$$

As the orientation is along the stator voltage vector and the supply voltage is balanced, $V^e_{qs} = 0$ and $V^e_{os} = 0$. It follows from Eqn (3.47) that

$$v_{as} = 338.8 \ \cos \omega t$$

$$v_{bs} = 338.8 \ \cos (\omega t - 120°)$$

and

$$v_{cs} = 338.8 \cos (\omega t + 120°)$$

From Eqn (3.46), for a squirrel cage induction machine

$$V_{dq}^e = [414.9 \ 0 \ 0 \ 0]^t$$

Hence, $\vec{V}_s = 338\angle 0$ is the voltage vector.

For the steady-state condition, $p = 0$ in the voltage equations (3.49)–(3.52). For $s = -0.025$, from Eqn (3.49)–(3.54), the steady-state impedance matrix is

$$Z_{dq}^e = \begin{bmatrix} R_s & -X_s & 0 & -X_m \\ X_s & R_s & X_m & 0 \\ 0 & -sX_m & R_r' & -sX_r' \\ sX_m & 0 & sX_r' & R_r' \end{bmatrix}$$

$$= \begin{bmatrix} 0.085 & -7.18 & 0 & -7 \\ 7.18 & 0.085 & 7 & 0 \\ 0 & 0.175 & 0.07 & 0.1795 \\ -0.175 & 0 & -0.1795 & 0.07 \end{bmatrix}$$

So the d-q axis currents are

$$I_{dq}^e = (Z_{dq}^e)^{-1} V_{dq}^e = [-141.08 \ -79.66 \ 145.68 \ 20.85]$$

Therefore, the stator and the rotor current vectors are

$$\text{Stator: } \vec{I}_s = \sqrt{141.08^2 + 79.66^2} \ \angle \tan^{-1} \frac{-79.66}{-141.08}$$

$$= 162\angle -150.55$$

$$\text{Rotor: } \vec{I}_r' = \sqrt{145.68^2 + 20.85^2} \ \angle \tan^{-1} \frac{20.85}{145.68}$$

$$= 147.16\angle 8.14°$$

From Eqn (3.47), the phase-a stator current

$$i_{as} = \sqrt{\frac{2}{3}} (-141.08 \cos \omega t + 79.66 \sin \omega t)$$

$$= 132.3 \cos(\omega t - 150.55)$$

Hence, the rms value of the stator current is

$$I_{as} = \sqrt{\frac{2}{3}} \frac{1}{\sqrt{2}} \sqrt{141.08^2 + 79.66^2}$$

$$= \frac{137}{\sqrt{2}} = 97 \, \text{A}$$

Likewise, the rms rotor current, referred to the stator, is

$$I'_{ar} = \sqrt{\frac{2}{3}} \frac{1}{\sqrt{2}} \sqrt{145.68^2 + 20.85^2}$$

$$= 84.96 \, \text{A}$$

$$P = \sqrt{3} \times 415 \times (132.3/\sqrt{2}) \cos 150.55^\circ$$

$$= -58.56 \, \text{kW}$$

$$\text{Check}: \quad P = V_d I_d + V_q I_q$$

$$= -414.9 \times 141.08 = -58.54 \, \text{kW}$$

The flux linkages, according to Eqn (3.53) and (3.54), are

$$\lambda_{ds}^e = (-22.85 \times 141.08 + 22.3 \times 145.68) \times 10^{-3}$$

$$= 0.0266 \, \text{Wb turns}$$

$$\lambda_{qs}^e = (-22.85 \times 79.66 + 22.3 \times 20.85) \times 10^{-3}$$

$$= -1.3553 \, \text{Wb turns}$$

$$\lambda_{dr}^e = (22.85 \times 145.68 - 22.3 \times 141.08) \times 10^{-3}$$

$$= 0.1827 \, \text{Wb turns}$$

$$\lambda_{qr}^e = (22.85 \times 20.85 - 22.3 \times 79.66) \times 10^{-3}$$

$$= -1.3 \, \text{Wb turns}$$

According to Eqn (3.57), the counter torque produced by generator action

$$T_g = \frac{P}{2} (\lambda_{ds}^e i_{qs}^e - \lambda_{qs}^e i_{ds}^e)$$

$$= \frac{6}{2}(-79.66 \times 0.0266 - 1.3553 \times 141.08)$$

$$= -580 \ \text{N m}$$

The flux linkage space-vectors are

$$\text{Stator:} \quad \vec{\lambda}_s = \sqrt{0.0266^2 + 1.3553^2} \ \angle \tan^{-1}\frac{-1.3553}{0.0266}$$

$$= 1.3553\angle -88.8°$$

$$\text{Rotor:} \quad \vec{\lambda}_r = \sqrt{0.1827^2 + 1.3^2} \ \angle \tan^{-1}\frac{-1.3}{0.1827}$$

$$= 1.312\angle -82°$$

Figure 3.26 shows the various space-phasors for orientation in the direction of the voltage vector. Note that the stator flux and the voltage vectors maintain a practically orthogonal relationship.

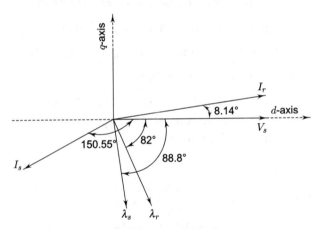

Fig. 3.26 Space-phasor diagram

(b) Referring to Fig. 3.26, if $\vec{\lambda}_s$ is taken as the reference space-vector, i.e., the stator flux direction is taken as the *d*-axis, the various *d-q* axis quantities become

$$\lambda_{ds} = 1.355 \ \text{Wb turns}$$

$$\lambda_{qs} = 0$$

$$\lambda_{dr} = 1.31 \cos 6.8° = 1.30$$

$$\lambda_{qr} = 1.31 \sin 6.8° = 0.155$$

$$I_{dr} = 147.16 \cos 96.94° = -17.78$$
$$I_{qr} = 147.16 \sin 96.94° = 146.08$$
$$I_{ds} = 162 \cos(-61.75°) = 76.07 \text{ A}$$
$$I_{qs} = 162 \sin(-61.75°) = -142 \text{ A}$$
$$V_{ds} = 338.8 \cos 88.8° = 7.1 \text{ V}$$
$$V_{qs} = 338.8 \sin 88.8° = 338.7 \text{ V}$$

3. A 3-ϕ, four-pole, 400-V, 250-kVA, 0.8-power-factor (lag), star-connected synchronous generator is driven through a gear box from a horizontal-axis wind turbine and delivers power to a unit-power-factor load, which is equivalent to a resistance of 1 Ω per phase. The following are some of the machine parameters in stator terms.

$$d\text{-axis mutual inductance, } L_{md} = 0.0021 \text{ H}$$
$$d\text{-axis stator self-inductance, } L_d = 0.0024 \text{ H}$$
$$q\text{-axis stator self-inductance, } L_q = 0.0015 \text{ H}$$
$$\text{Stator resistance, } R_s = 0.01 \text{ }\Omega$$

The open-circuit terminal voltage is 440 V when driven at 1500 rpm. If the machine runs at 1450 rpm, obtain the machine terminal voltage on load and the output power.

Solution

$$\text{Rotor speed } \omega_r = \frac{2\pi \times 1450}{60} \times \frac{4}{2}$$
$$= 303.7 \text{ elec rad/s}$$

From Eqn (3.75), for the open-circuit voltage (line-to-line) of 440 V at 1500 rpm,

$$\frac{\omega_e L_{md} i'_f}{\sqrt{3}} = \frac{440}{\sqrt{3}}$$

so,

$$i'_f = \frac{440}{2\pi \times 50 \times 0.0021}$$
$$= 667 \text{ A}$$

From Eqns (3.65), for a load resistance R_L and the steady-state condition ($p = 0$ and $I'_{dk} = I'_{qk} = 0$),

$$(R_L + R_s)I^e_{ds} - \omega_e L_q I^e_{qs} = 0$$

and

$$(R_L + R_s)I^e_{qs} + \omega_e(L_{md}I'_f + L_d I^e_{ds}) = 0$$

Using the parameters, we have

$$1.01 I^e_{ds} - 303.7 \times 0.0015 I^e_{qs} = 0$$

and

$$1.01 I^e_{qs} + 303.7(0.0021 \times 667 + 0.0024 I^e_{ds}) = 0$$

The solution to the above equations yields

$$I^e_d = -143.3 \, \text{A}$$

and

$$I^e_q = -317.8 \, \text{A}$$

The rms value of the phase current, from Eqn (3.80), is

$$I_a = \frac{\sqrt{143.3^2 + 317.8^2}}{\sqrt{3}}$$
$$= 201.3 \, \text{A}$$

The machine line voltage on load

$$V_{L-L} = \sqrt{3} I_a R_L$$
$$= \sqrt{3} \times 201.1 \times 1$$
$$= 348.6 \, \text{V}$$

The output power,

$$P_o = \sqrt{3} \times 348.6 \times 201.3$$
$$= 121.5 \, \text{kW}$$

The electromagnetic torque, according to Eqn (3.69), is

$$T_e = \frac{P}{2} \left[L_{md}I'_f I^e_{qs} + (L_d - L_q)I^e_q I^e_{ds} \right]$$

$$= \frac{4}{2} [-0.0021 \times 667 \times 317.8 + 143.3 \times 317.8$$
$$\times (0.0024 - 0.0015)]$$
$$= -807.4 \, \text{N m}$$

The input power,

$$P_{\text{in}} = (\omega_r T_e) \frac{2}{P}$$
$$= -\frac{303.7}{2} \times 807.4$$
$$= -122.63 \, \text{kW}$$

$$\text{Copper loss} = 3 \times 201.25^2 \times 0.01$$
$$= 1.21 \, \text{kW}$$

The output power,

$$P_o = 122.63 - 1.21 = 121.43 \, \text{kW} \tag{3.126}$$

Summary

This chapter contains the principle of operation and mathematical analysis of electrical machines to the extent required for their application in the conversion of wind energy to electricity. From wind turbines of only a few kilowatts capacity in the first decade of the twentieth century, technological developments have made it possible to reach turbine capacities of the order of 4.5 MW in a single unit. In the early days, dc generators were used, which still find application in low-voltage, low-capacity wind power systems charging storage batteries to operate lights and small appliances. For larger machines, dc machines have been phased out, mainly due to the problems associated with commutators. Ac generators, namely, induction and synchronous generators, are used by all major wind turbine manufacturers. Hence it is necessary to study ac generators in greater detail to understand their operation with wind turbines.

Starting with a brief introduction to the constructional features and rotating magnetic field, the chapter sequentially presents the derivation of the various circuit models for an induction machine, its torque and power relations, the saturation effect and its representation, and the control of rotor electrical power (slip power) either by controlling the rotor resistance or by injecting an emf in the rotor circuit.

It is expected that a wind turbine driven generator should match the high-quality power available from the utility. Constant-speed, high-capacity wind turbines with fixed-ratio gear boxes use squirrel cage induction generators. Variable-speed turbines, with some obvious advantages (presented in a later chapter), use wound rotor induction and synchronous generators, and permanent magnet ac generators. Owing to this present trend, this chapter presents the working principles of these machines to allow the reader to appreciate the applicability of these machines in various wind power generating schemes.

The concept of vector control is introduced and the dynamic d-q axis model of the induction machine is presented to facilitate the understanding of the induction generator's operation with a variable-speed turbine—which is dealt with in the later chapters. The chapter then presents the basic features of wound-field and permanent magnet synchronous machines, their mathematical models in a d-q reference, and the equations governing their steady-state operation. Equations governing the power flow between two synchronous sources are also given. The chapter closes with some worked out examples.

Problems

1. A four-pole, 400-V, 50-Hz squirrel cage induction motor has the following star-equivalent parameters referred to the stator: $R_s = 0.65\ \Omega$, $R'_r = 0.55\ \Omega$, $X_{\mathrm{ls}} = 1.52\ \Omega$, $X'_{\mathrm{lr}} = 2.28\ \Omega$, $X_m = 58\ \Omega$. The machine is connected to a 400-V, 50-Hz source and driven at 1545 rpm. Calculate (i) the power fed into the source, (ii) the developed electromagnetic torque, and (iii) the stator and rotor currents in a synchronously rotating reference frame with the d-axis aligning with the rotating stator mmf.

2. A four-pole permanent magnet three-phase machine with uniform air gap has the following parameters:

Resistance of each winding: 2.50 Ω

Leakage inductance of each winding: 1.50 mH

Mutual inductance between phase windings: 4.15 mH

The machine, when driven at 1500 rpm, gives an open-circuit line voltage of 110 V (rms). The machine is now connected to a balanced three-phase resistive load of 50 Ω per phase. Obtain the winding currents in the rotor reference frame, power delivered to the load, and the electromagnetic torque.

3. A three-phase, 50-Hz hydro generator is rated at 13.8 kV, 100 MVA, and 0.85 power factor (lag). The machine parameters at 50 Hz are the following:

d-axis stator self-reactance, X_d: 1.620 Ω

q-axis stator self-reactance, X_q: 0.915 Ω

Stator leakage reactance, X_l: 0.229 Ω

Stator resistance, R_s: 0.0036 Ω

The generator is working under rated conditions. At a certain instant the currents in the a-phase, b-phase, and c-phase are 2.09, 2.09, and -4.18 kA, respectively. Obtain the d-axis and q-axis stator linkages in the rotor reference frame.

4. A dc link at 500 V feeds 15 kW to a 400-V, 50-Hz bus of a utility at a displacement power factor of 0.95 through a sine PWM VSI and a step-up transformer (200/400 V line-to-line) with 4.5 mH inductance in each line between the transformer and the VSI. Find the amplitude modulation index of the control signal and its phase relative to the utility bus voltage.

5. A 150-kW turbine has a rated speed of 40 rpm. If the rated tip speed and the C_P-λ curve are assumed independent of the size of the rotor, what will the rated torque and the rated speed for a 250-kW turbine be? Further, if the same air-gap force density (kN/m^2) is assumed for the generators driven by the turbines, how will the rotor volumes of the two generators compare?

4

Power Electronics

Wind power is characterized by its stochastic nature. Wind speed changes continuously, and along with it the energy flow. A variable-speed wind machine is able to extract significantly more energy than a constant-speed machine. The grid connection of wind generators is essential to exploit their potential. However, the generated power, voltage, and frequency from a variable-speed wind machine changes with the wind speed. In such a situation, the most significant potential for the advancement of wind power technology lies in the area of power-electronics-controlled variable-speed operation. The developments in power electronics have not only increased the energy productivity but have also resulted in the quality control operation of both fixed- and variable-speed wind turbines. In order to fully appreciate the impact of this technology, it is necessary to know and understand the basic ideas, principles, and components of general power electronic conversion circuits, which are finding widespread application in wind electric power technology.

4.1 Power Electronics

In broad terms, the task of power electronics is to control the flow of electrical power efficiently by shaping the input voltage/current using power semiconductor devices so that the output voltage/current conforms to the load requirement. The system consisting of power semiconductor devices (and possibly other passive components) that performs this task is called a *power electronic converter*. The converter, in its turn, is controlled by integrated circuits. Thus, a power electronic converter controls and shapes an electrical input of magnitude V_i/I_i, frequency f_i, and number of phases m_i into an electrical output of magnitude V_o/I_o, frequency f_o, and number of phases m_o. The power flow through such a converter may be reversible, thus interchanging the role of the input and the output. This idea is presented in Fig. 4.1.

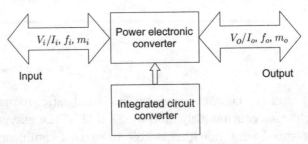

Fig. 4.1 Block diagram of a basic power electronic converter

4.1.1 Classification of Power Electronic Converters

There are many criteria for classifying converters used in power electronics. These include classification by the types of devices used, the function of the converters, the switching methods, etc. Unfortunately, there are no well-defined categories, based on these criteria, because there are always exceptions. In the following text, we will classify power electronic converters based on the nature of the input and the output, as shown in Fig. 4.2. We will show later how some of these converters are utilized in a power-electronics-controlled variable-speed wind turbine system.

4.1.2 Components of Power Electronic Converters

As has been mentioned earlier, power electronic converters consist mainly of power semiconductor devices operating in the switching mode, although other passive components, such as capacitors and inductors, may be used in some configurations. However, to avoid

Input Output	Ac	Dc
Dc	Rectifier	Dc–dc converter and choppers
Ac	Cycloconverter	Inverter

Fig. 4.2 Classification of power electronic converters based on the nature of the input and the output

power loss, resistors are never used in the main power path. In any case, power semiconductor devices are at the heart of all power electronic converters. In contrast to linear electronic systems, semiconductor devices in power electronic systems operate as switches, being either fully off or fully on.

An ideal switch should have the following characteristics.

(a) Bidirectional current conduction

(b) Bidirectional voltage blocking

(c) Zero on-state voltage drop

(d) Zero off-state current

(e) Zero switching time

(f) No power requirement for control

Obviously an ideal switch is not realizable. In power electronic converters, ideal switches are approximated by semiconductor devices with high switching speed, low on-state drop, and low off-state current. However, semiconductor switches are generally unidirectional.

Power semiconductor switches can be classified as shown in Fig. 4.3. In an uncontrolled semiconductor device, e.g., the diode or the diac, the state of conduction depends on the polarity of the applied voltage.

Conduction in a semi-controlled semiconductor switch is initiated by a control signal but terminated by reversing the polarity of the applied voltage. Thyristors and triacs are examples of such devices.

Fully controlled devices are turned on and off by the application or removal of a control signal. When the applied control signal

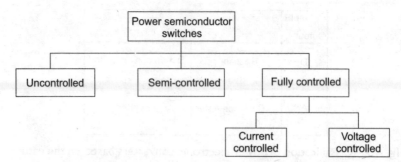

Fig. 4.3 Classification of power semiconductor switches

is current, the device is called a *current-controlled device* (e.g., BJT, GTO), and when the control signal is voltage, it is called a *voltage-controlled device*. The characteristics and application of these devices are discussed in the following sections.

4.2 Power Semiconductor Devices

4.2.1 Diodes

The simplest of all semiconductor devices, the diode is an uncontrolled switch. As shown schematically in Fig. 4.4(a), a diode is a *pn*-junction incorporated in a single silicon crystal. This junction is formed by diffusing, or alloying, a *p*-type impurity into an *n*-type silicon crystal. In a high-power diode the *n*-type region consists of a lightly doped region, called the *drift region*, grown on a heavily doped *n*-type substrate, as shown in Fig. 4.4(b). Figure 4.4(c) shows the circuit symbol of a diode. For conduction, the terminal A (the anode) must be at a positive potential with respect to the terminal K (the cathode). This is called *forward biasing*. In the forward-biased state, the diode is cut off until the potential V_{AK} reaches a threshold voltage V_{AF} of approximately 0.6 V. After this voltage is reached, the current through the diode increases very rapidly with V_{AK} as shown in the *I-V* characteristic of the diode in Fig. 4.4(d). The diode now behaves like an on switch. Under forward biasing, the voltage drop in the diode is of the order of 0.75 V at the rated current. The application of a reverse voltage (i.e., A negative with respect to K) causes a very feeble current to flow from the cathode to the anode. This is called the *reverse-biased* condition of the diode, and the junction withstands the reverse voltage applied. The diode now behaves like a switch that is turned off. If the reverse voltage exceeds a certain value, the junction breaks down. The drift region controls the reverse breakdown voltage. The reverse leakage current is of the order of a few milliamperes.

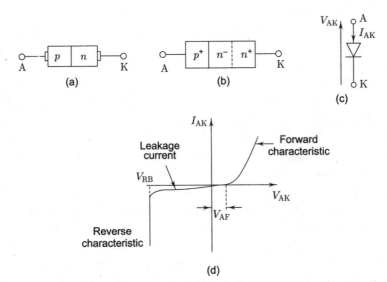

Fig. 4.4 Semiconductor diode: (a) schematic structure, (b) power diode structure with a drift layer, (c) circuit symbol, (d) I-V characteristic

A conducting diode, when turned off, needs a recovery time t_{rr} to attain its reverse blocking capability. Depending on the recovery characteristic, there are two basic types of diodes, namely, the general-purpose diode (converter grade) with t_{rr} of the order of 25 μs and the fast recovery diode (inverter grade) with t_{rr} less than 5 μs. The on-state voltage drop, the reverse breakdown voltage, and the reverse recovery time are considered the three important operating parameters in diode selection. General-purpose diodes are available with blocking voltage ratings of up to 5 kV and current ratings of up to several thousands of amperes. Inverter-grade diodes have ratings of up to 4.5 kV, 3.5 kA.

4.2.2 Thyristors

The thyristor, or the silicon-controlled rectifier (SCR), is a three-junction *pnpn* device with three terminals, as shown in

Fig. 4.5. This figure presents the cross section of a representative device showing the formation of an *npn–pnp* transistor equivalent circuit. Figure 4.5(b) shows its symbol and Fig. 4.5(c) presents its static *I-V* characteristics. A thyristor blocks current in both the directions in the off state. It can be triggered into conduction only in the direction from the anode (A) to the cathode (K) when the anode-to-cathode potential is positive and a positive current is injected into the gate relative to the cathode. It is a current-controlled device. When triggered into conduction, the device can remain on without any gate current once the device current reaches the latching current (I_L). When the current falls below the level of the holding current (I_H), the thyristor ceases to conduct. It regains the forward blocking state when it remains without any anode current and with a reverse voltage for a minimum period, known as the *turn-off time*. Based on this turn-off time, the thyristor can be of two types—the slow-speed phase control type with turn-off time of the order of 50–100 μs and the fast inverter type with turn-off time in the range 5–50 μs. The static characteristics further show that the increased gate current lowers the forward break-over voltage. Thyristors always operate below the forward break-over voltage (V_{FBO}) with no gate current.

Phase-control-type thyristors are generally used in line-frequency operated converters, and are available for voltages up to 4000 V and currents up to 3500 A. On the other hand, with inverter-grade thyristors, a maximum average current rating of 1500 A and a voltage rating of 2500 V have been achieved. The frequency of operation of a thyristor is generally limited to a maximum of 1 kHz.

Gate turn-off thyristor (GTO)

In a conventional thyristor the gate loses control after the device starts conduction. In ac circuits, reversal of the supply voltage causes commutation, and in dc systems some special commutation circuits are employed to turn the thyristors off. A gate turn-off thyristor (GTO) can be switched on or off by using the same gate terminal. Outwardly, the GTO has the same basic structure as the thyristor, but the requirement of the gate turn-off capability requires its architecture to differ. Figure 4.6(a) shows a simplified cross section of a GTO device. The width of the p_2-layer near the

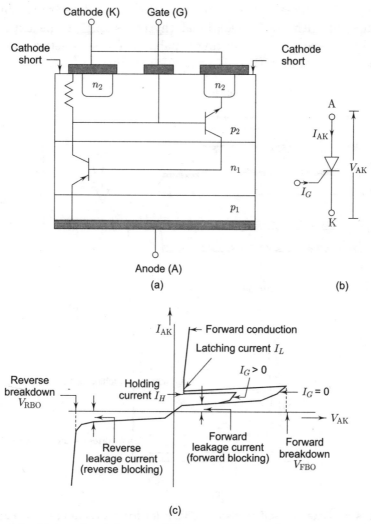

Fig. 4.5 Conventional thyristor: (a) basic structure with equivalent circuit, (b) device symbol, (c) static I-V characteristic with gate current effect

cathode is made smaller than in a conventional thyristor. The gate and cathode structures are made so as to maximize the periphery of the cathode and minimize the separation between their centres. In the anode region, the p-type layer is interrupted at regular intervals by highly doped n-spots (n^+) that make contact with both the n-base $(n_1$-layer) and the anode. This shorted anode structure reduces the turn-off and the forward recovery times compared with the conventional thyristors but at the expense of the reverse voltage blocking capability. For symmetric voltage blocking capability, there is no anode shorting.

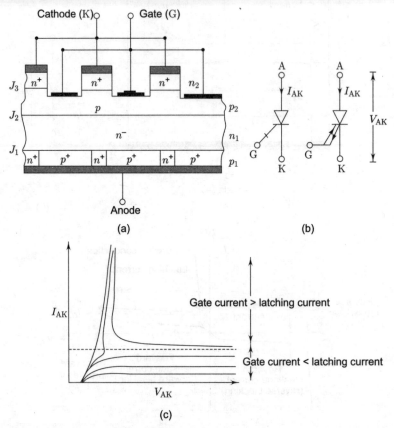

Fig. 4.6   Gate turn-off thyristor (GTO): (a) basic structure, (b) symbol, (c) I-V characteristic

Like the conventional thyristor, a GTO is turned on by injecting a positive current into the gate. However, due to the modified structure, the latching current demand is greater, and it is required to sustain the gate current at a level higher than the latching current to ensure latching. A small residual gate current is maintained during the on state to minimize the voltage drop. For any gate current lower than the latching value, the device behaves like a low-gain transistor. The circuit symbol for the GTO and the forward I-V characteristics are shown, respectively, in Figs 4.6(b) and (c). The GTO is switched off by injecting a high negative current pulse, relative to the cathode, into the gate. This negative gate current pulse is typically 20–35 % of the anode current. It removes the charges from the p-base (p_2-layer) faster than the rate at which they arrive at the base, and the junction J_2 can no longer remain in saturation. Junction J_3 becomes reverse biased and the collector current of $p_1 n_1 p_2$ is diverted by the gate, thus breaking the $p_1 n_1 p_2$–$n_1 p_2 n_2$ regenerative feedback effect. GTOs are available with voltage and current ratings of up to 4500 V and 3500 A with low turn-off times, in the range 5–15 µs. In spite of such a low turn-off time, the switching frequency of a GTO is limited to 2 kHz.

4.2.3 Bipolar Power Transistor

The bipolar power transistor, also called the bipolar junction transistor (BJT), is basically a three-layer n-p-n or p-n-p device with three terminals—the collector, the base, and the emitter, as shown in Fig. 4.7 for the n-p-n type. The base material is silicon. Commonly used transistors are now of the n-p-n variety because of the higher mobility of electrons. As such a transistor operates by the injection and collection of electrons as well as holes, it is called a bipolar transistor.

For better device switching and high-voltage properties, the collector region of a BJT is obtained by depositing an n-type drift layer on a highly conductive n^+ substrate using epitaxial growth. The p-base and the n^+ emitter regions are sequentially diffused into an epitaxy. The BJT is a current-controlled device and the base current I_B controls the collector current I_C. The ratio of the collector current to the base current is called the *current amplification factor*, or the *current gain*, β. The typical collector current versus output voltage characteristics with the base current as a parameter are shown in Fig. 4.7(c). In power switching applications, the transistor is operated in two states, which are referred to as the off state and the on state. In the absence of the base current, when $V_{BE} \leq 0$, the transistor is in the open-circuit condition. The collector–base junction supports the off-state voltage. For the transistor to act as turned on switch (very

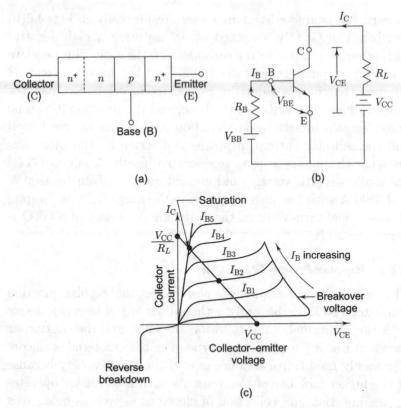

Fig. 4.7 The n-p-n transistor: (a) representative architecture and symbolic structure, (b) transistor circuit symbol, bias condition, and loading in the common emitter configuration, (c) output characteristics

small V_{CE}), a substantial base current must be injected to drive the transistor into the saturation state, as illustrated in Fig. 4.7(b). To ensure that the transistor actually operates in the saturated region, the base drive circuit is designed in such a way that

$$\frac{I_{C\,\max}}{I_B} \le \beta_{\min}$$

where $\beta_{\min}$ is the minimum current gain of the transistor (obtained from the data sheet). The reverse breakdown voltage of a power transistor is very low, as the base–emitter junction breaks down at a very low voltage, usually around 10 V, when V_{CE} is reversed. The device voltage and current ratings are limited to around 1500 V and 2000 A, respectively.

4.2.4 Power MOSFET

A power MOSFET (metal oxide semiconductor field effect transistor) is a unipolar majority carrier voltage-controlled device. As shown in Fig. 4.8(a), its output terminals are the drain and the source. The current from the drain to the source is controlled by controlling the gate voltage with respect to the source. Being a voltage-controlled device, the gate input impedance is very high, and essentially no gate input power is required to maintain the device in the fully on state. However, during fast turn-on and turn-off, low-power gate current pulses are needed to charge and discharge the input gate capacitance.

Normally off, n-channel enhancement type MOSFETs are used as switching devices. The DMOS-type structure of the device and its circuit symbol are shown in Figs 4.8(a) and (b), respectively. An n-type drift layer is grown on a highly conductive n^+ substrate. The p-body regions are next diffused into the n-regions, and the n^+-source regions are then diffused within the p-body regions. The gate structure is made with polysilicon, or other conducting materials, and embedded in the silicon dioxide insulating layer. The source contact metal spreads over the p-base and n^+-source regions. If the gate voltage is positive with respect to the source and exceeds a threshold value, a sufficient amount of positive charge is created on the metal gate, which, in turn, induces negative charges on the silicon surface beneath the gate oxide. This forms an induced n-layer in the p-region, thus providing a conductive path, i.e., an n-type channel, between the n^+-source region and the drain region. Current can flow from the drain to the source through this channel. As the gate voltage increases, the conductivity of this induced n-channel is enhanced and the drain-to-source current increases. As the source contact metal short-circuits the p-base region to the n^+-source, a parasitic diode is formed in the direction from the source to the drain. The output characteristics of a MOSFET are shown in Fig. 4.8(c). In power applications, the device is operated in the ohmic region to minimize the device voltage drop. The applied gate voltage will be such that the load current is below the saturation current corresponding to the applied gate voltage. Compared to the BJT, there are certain inherent advantages of the MOSFET, which

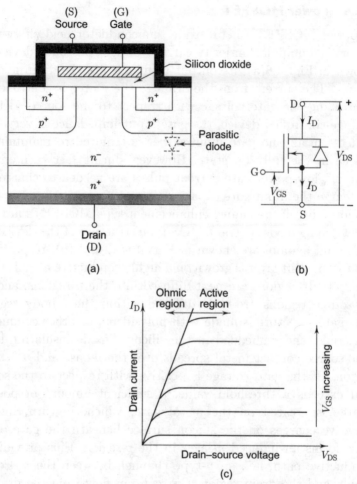

Fig. 4.8 The MOSFET: (a) basic structure, (b) device symbol, (c) output characteristics

include high switching speed (nanosecond order) resulting in low switching loss, very low control input power, the capability to operate over 100 MHz, the positive temperature coefficient of the on-state resistance, facilitating parallel operation of the devices, and high dV/dt immunity. On the other hand, high on-state resistance, increasing with the blocking voltage, and relatively low current and voltage capabilities compared to the BJT and other current-controlled devices are its limitations. Devices are available with ratings of up to 1000 V, 100 A.

4.2.5 Insulated Gate Bipolar Transistor

An insulated gate bipolar transistor (IGBT) is a hybrid transistor that combines the voltage-controlled operation and fast turn-on features of the MOSFET and the high power capability and low on-state voltage drop features of the BJT. Its turn-off time is related to the BJT operation. Figure 4.9(a) shows the basic architecture of an n-channel IGBT and the transistor structures embedded into it. Figure 4.9(b) shows the equivalent circuit and the device symbol. Structurally, an IGBT is similar to the MOSFET, except that the n^+n silicon of the MOSFET at the drain end is substituted by the p^+n silicon at the collector end. Owing to the introduction of the BJT operation, the load-side drain and source terminals are labelled as the collector and the emitter, respectively.

It is normally an off device. In the off-state, the junction J_2 blocks the forward voltage (collector positive with respect to the emitter), while the junction J_3 blocks the reverse voltage. With the collector at a positive potential, the application of a positive gate voltage beyond a threshold value, relative to the emitter, causes a current to flow, as in the MOSFET, through an induced n-channel in the p-base region under the gate. This is the base current for the p-n-p transistor. Consequently, current then flows from the collector to the emitter as the positively biased collector causes the p^+-region to inject a high concentration of minority carriers into the n-drift layer. It appears from the equivalent circuit that an IGBT may latch like a thyristor. However, the shorting of the p-base and the n^+-region avoids the turn-on of the parasite thyristor. Figure 4.9(c) shows the I-V characteristic with V_{GE} as the controlling parameter. The low reverse breakdown voltage is due to the n^+ buffer layer. V_{GE} is limited to 20 V. IGBT modules are available with ratings of up to 1700 V and 400 A and for switching frequencies of up to 100 kHz. The turn-on and turn-off delay times are typically in the ranges 55–320 ns and 250–800 ns, respectively.

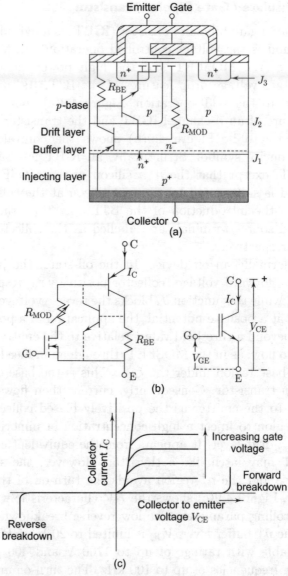

Fig. 4.9 *n*-channel IGBT: (a) simplified structure, (b) equivalent circuit and
device symbol, (c) output characteristics

4.3 Uncontrolled Rectifier

An uncontrolled converter is an unregulated rectifier employing
diodes to supply power to a dc circuit from an ac source. An

uncontrolled converter gives a fixed dc output voltage for a given ac supply. A single-phase rectifier is adequate for small power, whereas higher power levels generally use three-phase converter circuits.

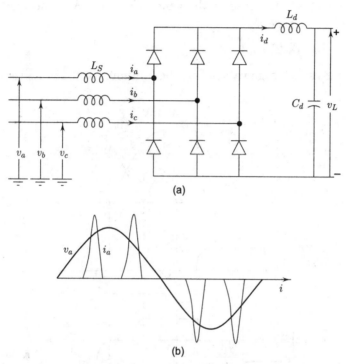

(a)

(b)

Fig. 4.10 Three-phase, full-wave uncontrolled rectifier: (a) circuit, (b) supply voltage and current waveforms with capacitive load

Figure 4.10(a) presents the circuit of a three-phase, full-wave diode bridge rectifier. In the circuit, L_d and C_d represent the filter on the dc side and L_S is the ac-side per-phase inductance, which may be the generator inductance or the line inductance of the supply system. The circuit behaviour depends on the values of L_d, C_d, and L_S.

If L_d and L_S are small and C_d is very large, practically smooth dc output voltage is obtained. The ripple content in the output voltage is very small and the ac-side current in each phase contains two spikes in each half-cycle, as illustrated in Fig. 4.10(b), introducing a high level of current harmonics in the ac source. In the absence of L_d and L_S, the current spikes will be very high. The inductances reduce the current surge and smoothen the supply-side current. Another unfavourable aspect of a diode-bridge rectifier with a large capacitor, when used directly with a synchronous generator, is the poor voltage regulation as a result of the increase in the generator load. With no load, the capacitor voltage is

$$V_L = \sqrt{2} V_{L\text{-}L}$$

For a given average value of i_d, the minimum value of L_d required

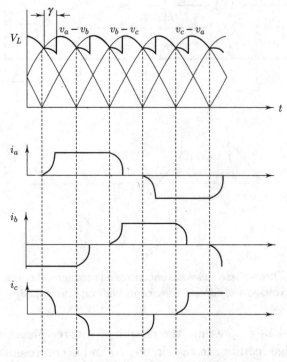

Fig. 4.11 Output voltage and phase current waveforms of a three-phase diode-bridge rectifier with inductive load

to make the current i_d continuous is given by

$$L_{d,\min} = \frac{0.013 V_{L\text{-}L}}{\omega I_d} \tag{4.1}$$

where ω is the supply frequency in rad/s.

If the inductor L_d is assumed to be very large, the diode-bridge output current will be an almost level current and the ac supply current in the absence of L_S will be symmetrical and have a quasi-square wave shape. However, in practice, the presence of L_S will cause the current through the outgoing diode to decrease gradually and the current through the incoming diode to build up gradually. The resulting phase current and the bridge output voltage waveforms are shown in Fig. 4.11. The overlap angle γ is given by

$$X_S I_d = \frac{V_{L\text{-}L}}{\sqrt{2}}(1 - \cos\gamma)$$

Overlapping of two diodes imparts a drooping characteristic to the average converter output voltage with increase in the load current. The average dc output voltage is

$$V_L = 1.35 V_{L\text{-}L} - \frac{3}{\pi} X_S I_d \tag{4.2}$$

Commutation should occur after every 60°. Equation (4.2) holds good for $\gamma \le 60°$. If the overlap angle exceeds 60° due to increase either in the load current or in the value of L_S, commutation between the two diodes in sequence will not commence until the ongoing commutation of the outgoing diode to the incoming diode is complete. This may cause the converter output voltage to have zero periods. This problem may arise with the diode-bridge rectifier when used with a synchronous generator having high values of reactance. Even with the drooping characteristic, voltage regulation for the same value of L_S is better with an inductive filter than with a capacitive filter. The use of a diode-bridge rectifier along with a synchronous generator having high values of reactances results in higher copper and iron losses, and instability when the diode-bridge rectifier is suddenly loaded. The drawbacks of a diode-bridge rectifier are overcome by using it along with a boost converter and a PWM rectifier, which are dealt with later.

4.4 Phase-controlled Converters

4.4.1 Line-frequency Naturally Commutating Rectifiers and Inverters

In many power electronics applications, it may be necessary to link the commercial sine wave ac source to a dc circuit to control the flow of power in one direction or in both directions at a regulated or unregulated voltage. This is accomplished by means of a converter. When the power flow is from the ac source to the dc side, the process is called rectification, but if the power flow is reversed, the converter is said to be in the inverting mode. A bidirectional converter that permits power flow in either direction is known as a full converter. All commercial power production is done by three-phase generators, and in industrial applications three-phase converters are preferred to single-phase converters because of the lower ripple content in the waveform, restriction on unbalanced loading, higher power handling capability, and reduced filter requirements.

Figure 4.12 shows a very commonly used circuit connection for a three-phase, line-commutated fully controlled converter, where each device is a thyristor. The voltage and current waveforms shown in Fig. 4.13 for continuous load current and zero source inductance $(L_s = 0)$ explain the steady-state operation of the circuit. The load current is assumed to be ripple-free. The output dc voltage is adjusted by exercising control over the firing delay angle α (Fig. 4.11) of the thyristors with respect to the ac voltage waveforms. The firing delay angle is measured from the point on the wave at which an ideal diode would naturally conduct or commutate. The triggering sequence is 1-2-3-4-5-6 and the turning-off of a conducting thyristor is brought about by the supply voltage reversal following the triggering of the next device in the sequence. A full-wave, three-phase diode bridge is a particular case with $\alpha = 0$. As the dc output voltage repeats after a duration of 60°, the mean output voltage for continuous load currents is given by

$$V_L = \frac{3}{\pi} \int_{\pi/6+\alpha}^{\pi/6+\alpha+\pi/3} v_{ab}\, d(\omega t) \qquad (4.3)$$

$$= \frac{3}{\pi} \int_{\pi/6+\alpha}^{\pi/6+\alpha+\pi/3} \sqrt{2} V_{L\text{-}L} \sin\left(\omega t + \frac{\pi}{6}\right) d(\omega t) \qquad (4.4)$$

$$= V_o \cos \alpha \qquad (4.5)$$

where $V_o = (3\sqrt{2}/\pi) V_{L\text{-}L}$.

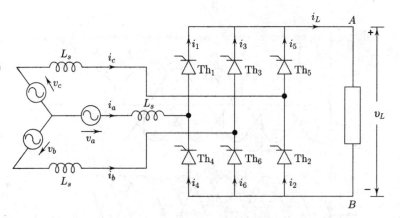

Fig. 4.12 Power circuit for a line-commutated, three-phase converter

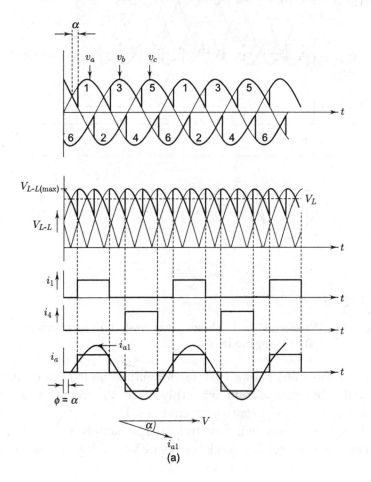

(a)

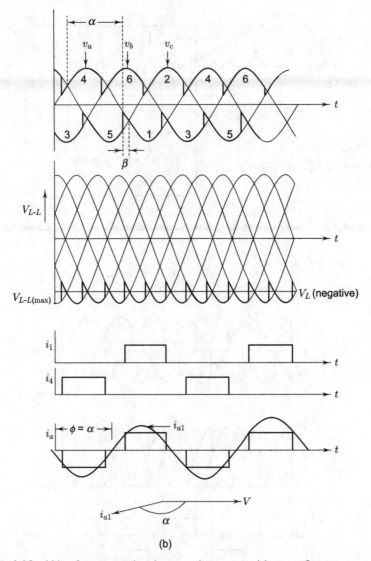

Fig. 4.13 Waveforms and phasor diagrams: (a) rectification mode, (b) inverting mode

Equation (4.5) shows that for a continuous load current and $L_s = 0$, changing α will smoothly vary V_L from its positive maximum to the negative maximum. For $\alpha < 90°$, V_L being positive, the power will flow from the ac supply to the dc side, indicating the rectifier mode of operation of the converter. For

$\alpha > 90°$, V_L becomes negative and the power flows from the dc side into the ac supply, representing converter operation in the inverting mode. This is illustrated in Fig. 4.13. As the current direction through a thyristor cannot be reversed, the dc terminal polarity changes for the reverse direction of the power flow. For this to occur, a dc source must be present, acting in the direction from A to B (Fig. 4.12). In practice, α is limited to approximately 150° owing to the source inductance and the thyristor turn-off time. In the inverting mode, the firing position of a thyristor is defined usually by the advance angle β with reference to the limiting value of α, i.e., 180°, as shown in Fig. 4.13(b). Then, α and β are related as follows:

$$\beta = 180° - \alpha \tag{4.6}$$

The derivation of Eqn (4.5) assumes an instantaneous transfer or commutation of current from one thyristor to the next. In fact, the presence of source inductance does not allow instantaneous collapse of current in a phase. As a result, current commutation is delayed. The outgoing and incoming thyristors conduct simultaneously till the current in the outgoing thyristor decays to zero, while the current in the incoming thyristor reaches the full load current. This interval is called the *overlap period*, and is specified by the angular period γ, called the *overlap angle*. γ is related to the converter output current I_d, the supply voltage $V_{L\text{-}L}$, the firing delay angle α, and the source reactance X_s $(= \omega L_s)$ by the following expression:

$$I_d X_s = \frac{\sqrt{3}}{2} V_{L\text{-}L} \left[\cos \alpha - \cos (\alpha + \gamma)\right]$$

The effect of the finite commutation interval is to reduce the mean dc output voltage, which can be shown to be

$$V_L = \frac{3\sqrt{2} V_{L\text{-}L}}{\pi} \cos \alpha - \frac{3 X_s I_d}{\pi} - 2V_t \tag{4.7}$$

where $2V_t$ is the total device drop.

Upon comparison with Eqn (4.5), it is seen that on loading, the mean open-circuit voltage $V_o \cos \alpha$ is reduced by $3X_s I_L/\pi + 2V_t$, giving the equivalent circuit shown in Fig. 4.14(a), representing the rectifier mode of operation of the converter.

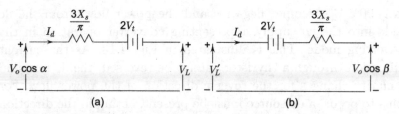

Fig. 4.14   Equivalent circuit of the converter on load: (a) rectification,
(b) inversion

For converter operation in the inverting mode, α is made greater
than 90°. In terms of the firing advance angle β of Eqn (4.6),
Eqn (4.7) becomes

$$V_L' = -V_L = \frac{3\sqrt{2}}{\pi}V_{L\text{-}L}\cos\beta + \frac{3X_L}{\pi}I_d + 2V_t \qquad (4.8)$$

The actual dc voltage direction in Fig. 4.13 is reversed during
inversion. Denoting this reversed voltage by V_L', the equivalent
circuit of the converter in the inverting mode is shown in
Fig. 4.14(b). The phase current waveform in Fig. 4.13 shows that
the firing delay causes a lagging displacement of the supply current
relative to the phase voltage. For continuous level load current,
with no overlap, the fundamental component of the ac supply
current lags behind the supply voltage by an angle ϕ, called the
displacement angle, equal to the firing delay angle α. Irrespective
of the mode of operation (rectification/inversion), the fundamental
current has a component lagging behind the supply voltage by
90° (Fig. 4.13). This implies that during the entire operation
$(0 \leq \alpha < 180°)$, a fully controlled converter must be supplied
with reactive volt-amperes (VAR) by the supply. This holds
good for any profile of the load current with line-commutated
converters. In the inverting mode of operation, the phase current
has a component in opposition to the voltage as is evident from
Fig. 4.13(b). This is due to the adoption of the rectifying mode as
the positive convention. When the dc side is the source of power
feeding the ac system and the current direction cannot be reversed,
the voltage polarity has to reverse compared to the rectifying mode
as shown in Fig. 4.14 for a positive value of power. This means
that, during inversion $(\alpha > 90°)$, power is fed back into the source
at a leading power factor.

4.4.2 Phase-controlled Ac Voltage Regulator

The phase-controlled ac voltage regulator is used to vary the effective output voltage at the same supply frequency, similar to what an autotransformer does. However, the output voltage deviates from a sinusoid. A number of circuits are available, of which the circuit most commonly used to control the rms output voltage for a three-phase system is shown in Fig. 4.15. In each line, two SCRs are connected in antiparallel to make a full-wave voltage controller. They are triggered cyclically at 60° intervals, similar to a phase-controlled rectifier/inverter. The control of the firing angle of the thyristors relative to the supply voltage determines the output voltage. This type of a circuit is used as a stator voltage controller for a squirrel cage induction motor to give it a soft start, and also to vary its speed under certain types of loads. Such circuits are also used as energy savers for lightly loaded induction motors. These converters cause heavy distortions of the supply current and have a poor line-side displacement factor.

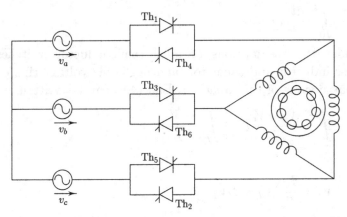

Fig. 4.15 The power circuit of an ac supply voltage regulator

4.5 Dc–dc PWM Converters

The dc output derived from a single-phase, or three-phase, diode-bridge rectifier with a capacitor filter or from a photovoltaic cell is usually unregulated. Dc–dc converters are used to obtain regulated dc output at a required voltage level from such unregulated dc supply. Of the many possible variations in converter topologies, we will describe two commonly used dc–dc PWM converter circuits, operated at high frequency.

4.5.1 Buck (Step-down) Converter

As the name implies, a buck (step-down) converter is used to obtain a regulated voltage less than the input voltage. Figure 4.16(a) shows the basic circuit of a buck converter using an IGBT switch. When the IGBT is switched on, energy from the supply flows into the inductor, capacitor, and the load. As a result, the inductor current rises. When the IGBT is turned off, the inductor current continues to flow through the freewheeling diode D_F, transferring some of its energy to the capacitor and the load. The inductor current falls until the IGBT T_d is switched on again. The current waveforms under continuous conduction of the inductor current are shown in Fig. 4.16(b). In the steady state, the total current change in the inductor over one switching period T_s is zero, i.e.,

$$\int_0^{T_s} \frac{di_L}{dt} dt = 0 \tag{4.9}$$

Under the assumptions of ideal semiconductor switches, a lossless inductor and capacitor, and negligible voltage ripple due to a large value of the capacitor, Eqn (4.9) can be written as

$$\int_0^{T_{ON}} \frac{V_d - V_o}{L} dt + \int_{T_{ON}}^{T_s} \frac{-V_o}{L} dt = 0$$

i.e.,

$$V_o = \frac{T_{ON}}{T_s} V_d = DV_d$$

where $D = T_{ON}/T_s$ is the duty ratio.

Thus, by varying the IGBT conduction duty ratio, the output dc voltage can be held at a fairly constant value even if the input dc voltage is unregulated. V_o is always less than the input voltage, and takes a different expression for the discontinuous mode of the inductor current. But even in that case, V_o can be kept constant by adjusting the duty ratio D.

In practice, the capacitor is of finite size, and charges and discharges, respectively, during T_{ON} and T_{OFF}. Consequently, ripples

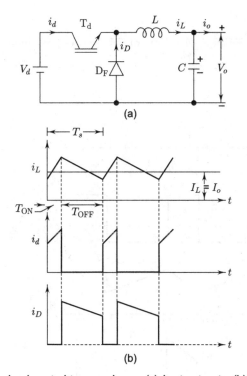

Fig. 4.16 The buck switching regulator: (a) basic circuit, (b) current waveforms with continuous inductor current

appear in the output voltage. These ripples can be minimized to less than 1% by selecting the values of L and C in such a way that $1/2\pi\sqrt{LC}$ is much less than the switching frequency f_s $(= 1/\tau_s)$ of the IGBT. For high switching frequencies in the kilohertz range, MOSFETs and IGBTs are used in such PWM converters.

4.5.2   The Boost (Step-up) Converter

The boost converter steps up the input voltage. Figure 4.17(a) presents the basic circuit. When the transistor is turned on, the source voltage appears across the inductor, and the inductor current builds up. The diode D becomes reverse biased and remains off. When the transistor is turned off, the inductor releases its energy to the capacitor and the load via the diode D. The inductor current decreases. It replenishes the loss of charge on the capacitor which occurs during the on period of the transistor in order to maintain the load current.

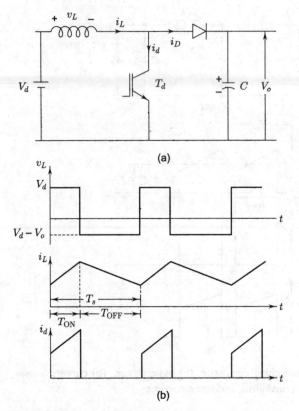

(a)

(b)

Fig. 4.17 Dc–dc Boost converter: (a) power circuit, (b) steady-state wave-forms

The waveforms shown in Fig. 4.17(b) relate to steady-state operation of the circuit under the continuous current mode of the inductor. The condition of zero net change in the inductor current over a period of steady-state operation gives

$$\int_0^{T_s} \left(\frac{di_L}{dt} \right) dt = 0$$

i.e.,

$$\int_0^{T_{ON}} \frac{V_d}{L} dt + \int_{T_{ON}}^{T_s} \frac{V_d - V_o}{L} dt = 0 \qquad (4.10)$$

Taking the capacitor to be very large, there will be no significant ripple in the output voltage. Under the assumption of constant output voltage and continuous inductor current, Eqn (4.10) yields the following relation between the output voltage and the input voltage:

$$\frac{V_o}{V_s} = \frac{1}{1 - D}$$

where $D = T_{ON}/T$ is the duty cycle of the transistor.

The continuous current in the inductor will persist if

$$L > \frac{(1 - D)^2 DR}{2f}$$

To limit the output voltage ripple to δV, the minimum value of the capacitor should be

$$C_{\min} = \frac{DV_o}{(\delta V_o)Rf}$$

4.5.3 Buck–Boost Converter

This converter, shown in Fig. 4.18(a), provides an output voltage which may be less than or greater than the input voltage. When the IGBT T_d is turned on, the diode becomes reverse biased and the input energizes the inductor, increasing its current. When the IGBT is turned off, the inductor current is forced through D_F, and a part of the inductor energy is transferred to the capacitor and the load. The inductor current falls until the IGBT is turned on again, and the next cycle commences. The converter provides negative polarity output voltage with respect to the input. Figure 4.18 shows the steady-state current waveforms for continuous inductor current. In the steady state, waveforms repeat and the total change in the inductor current over one cycle is zero, i.e.,

$$\int_0^{T_s} \frac{di_L}{dt} dt = 0 \tag{4.11}$$

Assuming the capacitor to be large enough to keep the output voltage V_o almost constant, the integral in Eqn (4.11) can be evaluated as

$$\frac{V_d}{L} T_{ON} - \frac{V_o}{L}(T_s - T_{ON}) = 0$$

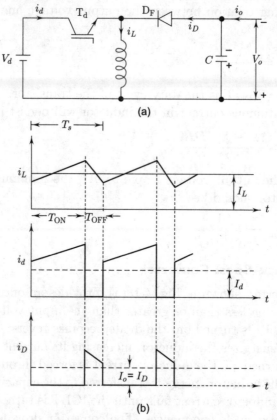

(a)

(b)

Fig. 4.18 Dc-to-dc buck–boost switching regulator: (a) basic circuit, (b) current waveforms with continuous inductor current

Thus,

$$V_o = \frac{T_{ON}}{T_s - T_{ON}} V_d \qquad (4.12)$$

$$= \frac{D}{1 - D} V_d \qquad (4.13)$$

where $D = T_{ON}/T_s$ is the duty ratio of the switch. For $D < 1/2$, the output voltage is less than the input voltage, while for $D > 1/2$, the output voltage is greater than the input voltage.

In actual operation, the output voltage fluctuation cannot be avoided, because of the discharging and charging of the capacitor during the on and the off periods of the switch, respectively. For reduced fluctuation of the output voltage, the switching period T_s

of the transistor switch should be selected to be much lower than the time constant τ $(= CV_o/I_o)$ of the output circuit.

As can be seen from the current waveforms in Fig 4.18(b), the input current is always discontinuous. This is overcome in the Cúk converter, shown in Fig. 4.19, which has a continuous input current. It is another circuit topology of the buck–boost converter, and the input–output voltage relation is given by Eqn (4.13). Here the capacitor C_1 is the link through which energy is transferred from the input to the output circuit. When T_d is off, the diode D_F conducts. The capacitor C_1 receives energy from both the input and the inductor L_1, and the inductor current i_{L_1} decreases. The stored energy in L_2 feeds the load. When T_d is turned on, the voltage of the capacitor C_1 reverse biases the diode D_F. The capacitor releases some of its energy via T_d, L_2, C_2, and the load. The current in L_2 increases, and the input supplies energy to L_1, causing its current i_{L_1} to increase.

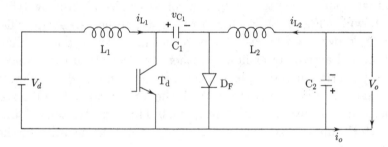

Fig. 4.19 Configuration of the Cúk regulator

4.6 The Inverter: Dc to Ac Conversion

Inverters are required to transfer energy from a dc source to an ac source, or load, at a desired voltage and frequency. Wind generators produce power at voltages and frequencies that generally, do not meet the requirements of consumers. In many wind farms, they are used to feed power into the utility grid from an unregulated dc bus formed from the rectification of the ac power of the wind generator. Ideally, an inverter should convert dc input voltage into sinusoidal ac output whose magnitude and frequency can be controlled. In actual practice, inverters produce rectangular output waveforms which contain harmonics. Special switching techniques are adopted to significantly reduce the effects of these harmonics. Except in small power applications, inverters with three-phase output are used, particularly in power generation schemes.

Inverters are of two types, namely, voltage source inverters (VSIs) and current source inverters (CSIs). A VSI is fed by a stiff dc voltage source—a source that is supposed to remain unaffected by the load, i.e., a source ideally with no internal impedance. A CSI receives constant direct current at the input in the presence of ac load impedance variation, and the terminal voltage changes. Ideally, the CSI sees a dc source at the input with an internal impedance approaching infinity.

4.6.1 The Basic Structure of a VSI

The most frequently used circuit of a three-phase, full-bridge voltage source inverter using controlled semiconductor switches is shown in Fig. 4.20. Each controlled device has an antiparallel diode, known as a bypass diode, to permit reactive power circulation and active power flow in either direction. Inverters usually take power from a dc link fed by a diode/controlled rectifier, a dc–dc converter, or a battery. With the VSI, a capacitive filter is placed at the inverter input terminals. The term 'voltage source inverter' implies that over a period of the output ac voltage, the

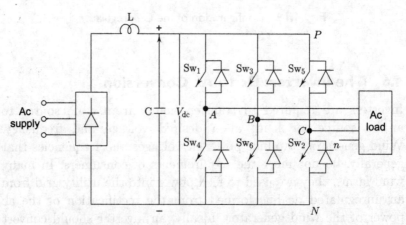

Fig. 4.20 Three-phase voltage source bridge inverter

input dc voltage is constant or shows negligible variation. The switching schemes for the three legs of the bridge are identical and the triggering signals are displaced at a phase difference of 120° to produce a balanced three-phase voltage at the output terminals. Depending upon the switching scheme, the inverters produce either quasi-square waveforms or pulse width modulated waveforms.

Quasi-square wave switching scheme

In one of the quasi-square wave switching schemes, the two switches in each of the three legs operate as a complementary pair. This means that each switch in a leg conducts for one half-cycle (180°) of the desired output frequency. Thus each output terminal switches every half-cycle between V_{dc} and 0 V with respect to the negative dc bus. This operation produces quasi-square wave line voltages and six-step phase voltages as shown in Fig. 4.21. The rms value of the line voltage is $\sqrt{2/3}V_{dc}$, or $0.816V_{dc}$, while the rms value of the fundamental component is $\sqrt{6}V_{dc}/\pi$, i.e., $0.78V_{dc}$. The inverter output voltage is controlled by adjusting the dc-link voltage, while the output frequency is controlled by changing the time interval between the consecutive control signals to the devices of the same phase leg. The line voltage as well as the phase voltage

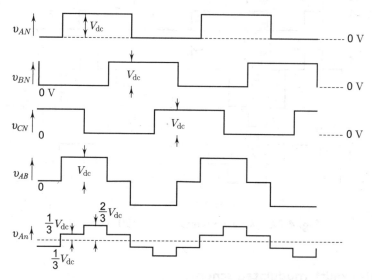

Fig. 4.21 Voltage waveforms with the 180° triggering scheme

contain harmonics of the order $6n \pm 1$, where n is any positive integer.

In another inverter control scheme, each controlled switch is kept on for 120° of the output cycle with 60° dead time between the triggering instants of the switches in the same leg. With the negative bus of the dc link as the reference, the potential variations of the output terminals will be as shown in Fig. 4.22. The line voltages will be six-step waveforms whereas the phase voltages will be quasi-square waves. These waveforms will be modified, approaching those shown in Fig. 4.21, if any inductance is present in the load circuit. When the controlled switch is turned off, the phase current will be transferred to a bypass diode, effectively keeping the inverter output terminals connected to the dc bus for more than 120°.

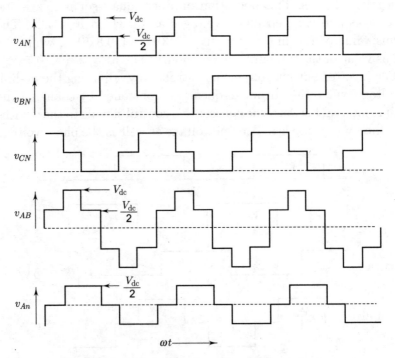

Fig. 4.22 Voltage waveforms for 120° conduction

Pulse width modulated scheme

The problems associated with the six-step and quasi-square wave inverters involve low-order harmonics (5th and 7th) and control of inverter output voltage to cope with the variation in the un-controlled dc input voltage. The requirement of ac output voltage with minimum distortion and filtering requirements along with the facility for output voltage control, without using any additional external controllable power device, can be met using a technique known as pulse width modulation (PWM). The inverter circuit remains the same as in Fig. 4.20, but the switches are turned on and off many times in a cycle to minimize low-order harmonics in the output voltage. In PWM techniques, constant-amplitude pulses are generated, and the pulse durations are modulated to obtain the desired shapes and magnitudes of the output. Several PWM techniques have been proposed, of which the sine-weighted PWM is commonly used. This method is illustrated in Fig. 4.23.

The three-phase sinusoidal PWM inverter implements a control strategy in which a high-frequency symmetrical triangular carrier wave is compared with three reference, or modulating, sine waves v_A^*, v_B^*, and v_C^* of desired frequency and amplitude, as shown in Fig. 4.23. The sine waves are equal in magnitude and frequency and phase-shifted from one another by 120°, forming a balanced system. If the carrier to the modulating wave frequency ratio, called the *frequency-modulation ratio* m_f, is a multiple of 3, the carrier wave has the same phase relation with each of the sine waves. This ensures identical phase voltage waves in the three-phase system.

The states of the inverter switches are decided by the relative magnitudes of the triangular wave and the sine waves. When the instantaneous sine wave of a phase exceeds the carrier wave voltage, the upper device in the leg of that phase is switched on. When the voltage of the reference wave falls below that of the triangular wave the lower device is turned on. It is the carrier frequency with which the inverter switches are operated, and the switching loss is greater in a PWM inverter than in a square-wave inverter. With reference to the negative side of the dc source in Fig. 4.20, the phase voltage terminals exhibit identical notched waveforms mutually displaced by 120°, as shown in Fig. 4.23, with pulse widths varying sinusoidally. The fundamental frequency equals that of the reference sine wave. The resulting line voltage waveforms are as shown in Fig. 4.23. The amplitude of the fundamental frequency component of the voltage (phase or line) is decided by the ratio of the sine wave amplitude V_m to the carrier

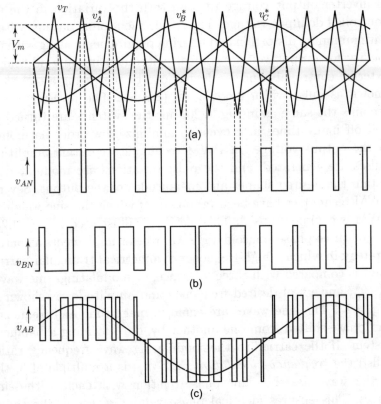

Fig. 4.23 Sinusoidal pulse width modulation in a three-phase voltage source inverter: (a) reference signals and carrier waveforms, (b) output voltage waveforms of the terminals A and B with respect to the negative bus, (c) line-to-line voltage waveforms (between terminals A and B)

wave amplitude V_t, termed the *modulation index* m_a, and the dc bus voltage. The line-to-line rms voltage at the fundamental frequency can be shown to be given by

$$V_{L\text{-}L1} = \frac{\sqrt{3}}{\sqrt{2}} m_a \left(\frac{V_{dc}}{2} \right) \tag{4.14}$$

$$= 0.612 m_a V_{dc} \tag{4.15}$$

where $m_a = V_m/V_T$. This linear relationship holds good for $m_a \leq 1$. Thus, for fixed amplitude of the triangular carrier wave, the fundamental frequency output voltage can be controlled by varying the sine wave amplitude, retaining the sinusoidal pattern in the output voltage waveform.

If m_a is increased beyond unity by allowing the amplitude of the reference sine wave to exceed the peak of the triangular wave, the fundamental rms does not increase proportionally with m_a, but depends on the frequency modulation ratio m_f. For a very large value of m_a, the sine wave PWM operation degenerates into the quasi-square wave operation, with the maximum value of the fundamental frequency line-to-line voltage equal to $0.78V_{dc}$.

The analysis of a three-phase system shows that if m_f is odd and a multiple of 3, the positive and negative output half-cycles are symmetrical, and no even harmonics are present in the output voltage. Triple frequency-related harmonics are also eliminated from the line voltage. The odd harmonics cluster as sidebands around the carrier frequency. With a large value of m_f, the high-order harmonics will dominate and can be more easily filtered than the low-order harmonics. Moreover, nominal line inductances will drastically attenuate these high-frequency harmonics in the current waveform. The output current waveform will be almost sinusoidal.

4.6.2 Current-regulated Pulse Width Modulation

Pulse width modulation methods involve the synchronization of the reference waves with the desired output waveforms, and a high degree of precision in generating waveforms. While feeding power into the utility network through the dc-link inverter, the output current and its phase become the control variables with regard to power transfer at constant grid voltage. Hence, it is desirable for the PWM control to be associated with the system current control. Two current controllers for voltage-fed PWM inverters, namely, the hysteresis controller and the ramp comparison controller, are in wide use. One does not require any information about the system parameters in order to use these controllers.

In both the methods, three-phase reference sine waves proportional to the desired currents, based on certain criteria, are generated and compared with the measured instantaneous values of the output currents. To keep the deviation within the prescribed limits, the comparator error is then processed through a controller generating signals to turn on or turn off the appropriate switching devices of a conventional PWM VSI.

Figure 4.24(a) shows the block diagram of the basic control and the power circuit. Figure 4.24(b) illustrates the principle of the hysteresis-band current control. The hysteresis comparator has a dead band that permits a deviation of Δi of the actual current from the reference wave. If an actual phase current exceeds the current reference by the hysteresis band, the upper device in the inverter leg of that phase is turned off and the lower device is turned on. This causes the phase current to decay until the current error reaches the lower limit when the switch status is reversed, causing the current to rise again. Independent control of phase current is possible in this manner if the three-phase load neutral is connected to the dc bus midpoint. In a system without such a neutral connection, the current response of a phase depends not only on the switching state of the corresponding leg but also on the status of the other two inverter legs. For loads with unconnected neutrals, the instantaneous current error can take double the hysteresis band.

The inverter switching losses are proportional to the switching frequency, which is influenced by the hysteresis band, the current level, the dc-link voltage, the counter emf, and the line reactances. Under favourable conditions the inverter switching frequency may become excessive, or vary over a range, producing objectionable acoustic noise. Fixed switching frequency operation ensures removal of the acoustic noise and prediction of the switching losses.

Figure 4.25 shows the fixed-frequency current control technique by the ramp comparison method for one leg of the inverter. Here the current error is compared with a fixed frequency triangular waveform to produce a PWM signal with a duty cycle proportional to the current error. If the error is positive, the on period of the upper device in the leg will be more than that of the lower device, increasing the ac line current in the positive direction. The operation resembles the asynchronous sine-triangle PWM with the current error as the modulating signal. Inherent magnitude and phase errors in the line current are reduced by adjusting the triangle amplitude, amplifying the current error, or adding compensation.

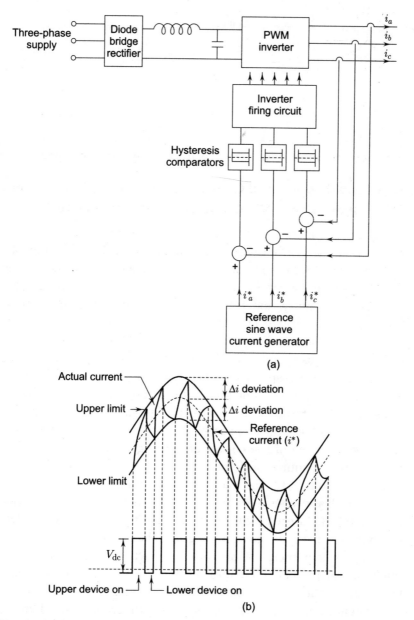

Fig. 4.24 Hysteresis-band current control: (a) block diagram of the three-phase PWM inverter with a hysteresis band current controller, (b) inverter output current and voltage waveforms with sinusoidal reference current

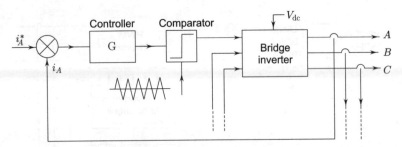

Fig. 4.25 Fixed-frequency current controller using pulse width modulation

4.6.3 Bidirectional and Controlled Power-factor Operation of the Sine-PWM Converter (Rectifier/Inverter)

Retaining the methodology described in the previous section to generate the triggering signals for switching devices on and off, the VSI of Fig. 4.20 can be made to work as a controlled converter through which power is reversible and which is also capable of working at unity and leading power factors. Two distinctive features in the bridge circuit of Fig. 4.26 are the inductances L on the ac side and a large capacitance on the dc side. The capacitance is used to ensure fairly constant voltage over a short period of time, irrespective of the transients and the switching events in the converter. The phase circuit inductance L is intended to assist in the indirect control of the current at a chosen power factor, thus helping in the process of selecting the operating mode (rectifier/inverter) of the converter. During inversion, the dc side must contain a source of dc power.

With reference to the rectifier convention in the circuit of Fig. 4.26(a) and considering the fundamental frequency components only, the steady-state ac-side quantities are related in the manner shown in the phasor diagrams of Figs 4.26(b) and (c). For the rectifier mode of operation, shown in Fig. 4.26(b), the fundamental frequency component V_I at the ac input terminals lags behind the source voltage V_s by an angle δ, and I has a component in phase with V_s. In the phasor diagram of Fig. 4.26(c), the voltage V_I leads V_s when the ac supply current I has a component which is in phase opposition to V_s, implying the inverting mode of operation, where the power flow is from the dc side to the ac side. In either case, neglecting the resistance of the inductor coil, the power flow through the inductor coil is given

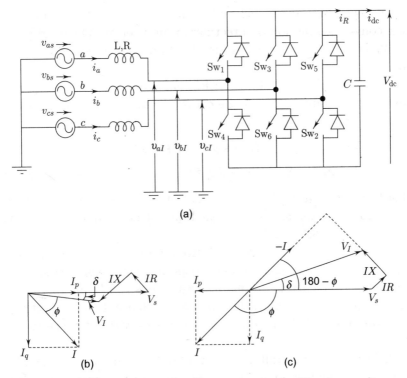

Fig. 4.26 Converter operation: (a) power circuit. (b) phasor diagram, rectifier mode, (c) phasor diagram, inverter mode

by

$$P = 3\frac{V_s V_I}{X_L} \sin \delta \qquad (4.16)$$

where δ is known as the load angle and is supposed to be positive for lagging phase.

Therefore, the current magnitude, power transfer, and the mode of operation (rectifier/inverter) can be controlled by adjusting the magnitude and/or phase (lag or lead) of V_I in relation to the ac supply voltage V_s.

Note from the waveforms of Fig. 4.23 that in the sine PWM inverter, the fundamental components of the ac terminal voltages of the inverter are in phase with the respective modulating sine waves. Hence, it is obvious that if the modulating sine waves, i.e., the control voltages, are phase delayed, or advanced, by an angle δ with reference to the ac supply voltage V_s, the converter ac terminal voltage will be delayed, or advanced, accordingly, making the converter function as a rectifier or an inverter. The magnitude of V_I is controlled by adjusting the amplitude modulation ratio m_a.

From the phasor diagram in Fig. 4.26(b) the in-phase and the quadrature components of the converter ac terminal voltages are found to be

$$V_I \cos \delta = V_s - RI \cos \phi - XI \sin \phi \qquad (4.17)$$

$$V_I \sin \delta = XI \cos \phi - RI \sin \phi \qquad (4.18)$$

For the inverting mode, with the current I flowing into the source at a power-factor angle ϕ_1, ϕ in Eqns (4.17) and Eqn (4.18) is replaced by $(180° - \phi_1)$.

V_s, R, and X are obtained from direct measurements. To operate the converter at a desired power-factor angle ϕ for any demand of the current I, i.e., of power, Eqns (4.17) and (4.18) are used to provide the values of V_I and δ, which decide the magnitude and the phase of the modulating sine waves with reference to the supply voltage.

Inverter terminal voltages v_{aI}, v_{bI}, and v_{cI} can be controlled by a vector controller for their magnitudes and required phase shift from the supply to meet the required flow of the active and the reactive powers. The voltage equations for the line inductors in Fig. 4.26 are

$$v_{abc,s} = Ri_{abc} + L\frac{di_{abc}}{dt} + v_{abc,I} \qquad (4.19)$$

In the d^e-q^e reference frame (Fig. 4.27), rotating at the supply angular frequency ω_e with the d^e-axis overlapping the supply voltage vector, the above voltage equations, from Eqn (3.46), become [cf Eqns (3.49) and (3.50)]

$$v_{ds}^e = Ri_d^e + Lpi_d^e - L\omega_e i_q^e + v_{dI}^e, \qquad (4.20)$$

$$0 = Ri_q^e + Lpi_q^e + L\omega_e i_d^e + v_{qI}^e, \qquad (4.21)$$

where v_{dI}^e and v_{qI}^e are the components of the inverter ac terminal voltage vector along the d^e- and q^e-axis, respectively. From Eqn (3.62),

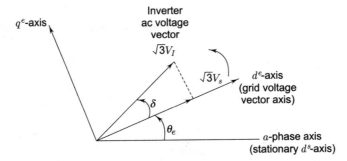

Fig. 4.27 The angular relationships

$$v_{ds}^e = \sqrt{3}V_s \tag{4.22}$$

(since $v_{qs}^e = 0$). From Eqn (3.63), the active power flow is

$$P = v_{ds}^e i_d^e \tag{4.23}$$

Using Eqns (4.21) and (4.22) in Eqn (4.23) for steady-state operation, i.e., $p = 0$, and neglecting R.

$$P = -\frac{\sqrt{3}V_s V_{qI}^e}{\omega_e L} \tag{4.24}$$

With reference to the relative positions of the voltage vectors and their components along the d^e-q^e axes in Fig. 4.27, the power flow given by Eqn (4.24) becomes

$$P = -\frac{3V_s V_I}{X_L}\sin\delta \tag{4.25}$$

which is the same as Eqn (3.113) except the sign implying power flow from the supply side to the inverter side.

From Eqn (3.61), the reactive power flow is

$$Q = -v_{ds}^e i_q^e \tag{4.26}$$

The use of Eqns (4.20) and (4.22) with voltage vector components along the d^e-q^e axes in Fig. 4.27 gives

$$Q = \frac{3V_s^2}{X_L} - \frac{3V_s V_I}{X_L}\cos\delta \tag{4.27}$$

which agrees with Eqn (3.114) except the sign because of the reverse direction of the power flow.

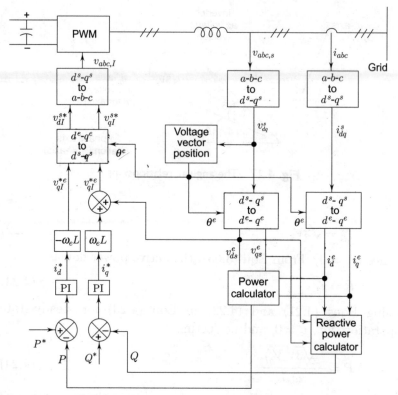

Fig. 4.28 Control scheme to regulate power flow

The vector control scheme to regulate the power flow is detailed in Fig. 4.28. The utility system voltage $v_{abc,s}$ is transformed into stationary d^s-q^s components by Eqn (3.48). As the d^e-axis of the synchronously rotating reference frame is aligned along the supply voltage vector, the angular position of the d^e-axis is computed as

$$\theta_e = \arctan \frac{V_q^s}{V_d^s} \tag{4.28}$$

The actual active and reactive powers, computed by Eqns (4.23) and (4.26), respectively, are compared with the reference values P^* and Q^* to generate the reference values of the inverter terminal voltages $v_{dI}^e{}^*$ and $v_{qI}^e{}^*$ in accordance with Eqns (4.20) and (4.21). These are then sequentially processed to generate the signals for the PWM inverter, forcing the inverter ac terminals to take values with the required phase shift from the supply for the necessary power flow.

4.6.4 Current Source Inverter

A current source inverter (CSI) is fed by a dc source whose current output is little affected by the inverter operation over a few cycles. A current source is obtained by connecting a large inductance in series with a dc voltage source, as shown in Fig. 4.29. The voltage imbalance at any instant between the dc source and the dc terminals of the inverter appears across the inductor, and is Ldi/dt. For a very large inductance L, di/dt will be very small, implying an effectively constant value for the current. With a change in the inverter power demand, the dc current level should ultimately settle to a new value. Alternatively, the dc source voltage may be adjusted to maintain the same current.

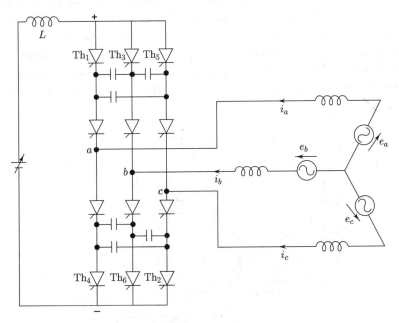

Fig. 4.29 Power circuit for the three-phase CSI

Figure 4.29 shows the power circuit diagram of a CSI, employing thyristors, with an active load in the output circuit. The active load may be the utility grid, a synchronous motor, or an induction motor. The devices are triggered in the sequence 1-2-3-4-5-6 at intervals of 60°. Ignoring the commutation overlap, each device remains on conduction for 120°. At any instant, only two devices conduct—one from the upper group and one from the lower group. For level input current and balanced sinusoidal emfs e_a, e_b, and e_c, the inverter terminal voltages will be nearly balanced sinusoids. Figure 4.30 shows the idealized waveforms of the inverter terminal voltages and the phase-a current for a triggering angle α. The fundamental current component lags behind the voltage by an angle ϕ that equals the firing delay angle α.

For the active power to flow from the dc side to the ac side, i.e., for the inverting mode, the firing delay angle α, which also gives the phase angle ϕ, must lie between 90° and 270°, as revealed by the phasor diagrams in Figs 4.30. Only then can the fundamental component of the load current have a component in phase opposition to the terminal voltage. Referring to the waveforms shown in Fig. 4.30(a) for phase a, with α in the range 90°–180°, the device Th$_4$ at the end of its 120° conducting period is commutated by triggering the forward-biased Th$_6$, which imposes negative v_{ab} across Th$_4$. This is line-commutated or load-commutated inverter operation. In this instance the inverter receives lagging VAR

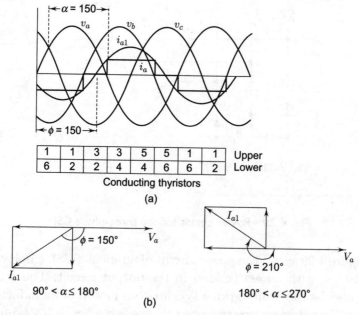

Fig. 4.30 (a) Ac-side voltage and phase-a current waveforms; (b) phasor diagrams for $90° < \alpha \leq 180°$ and $180° < \alpha \leq 270°$

(reactive volt-amperes) from the load. Conversely, the inverter supplies leading VAR to the load.

If the firing delay angle is adjusted between 180° and 270°, the voltage and the current waveforms and the phasor diagram for the fundamental components will be as shown in Fig. 4.31(b). The inverter operation still persists, as explained earlier. Two differences, with respect to the earlier inverter operation, are noted. At the end of its conducting period of 120°, the outgoing thyristor Th_4 is impressed with positive v_{ab} and hence can be turned off only by forced commutation. This is called *forced-commutated inverter operation*. The phasor diagram indicates that the load, in this case, supplies leading VAR to the inverter, i.e., lagging VAR flows to the load.

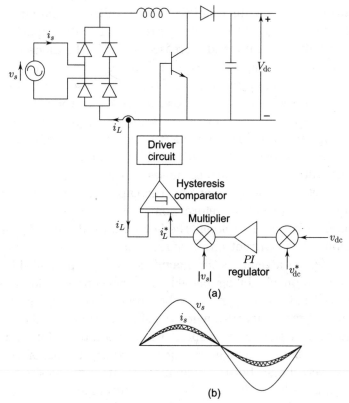

(a)

(b)

Fig. 4.31 Diode rectifier with boost chopper for current shaping, and the basic control scheme

4.7 Diode Rectifier and Input Line Current Shaping

The ac/dc/ac scheme is widely used to interface a wind generator with a three-phase utility source or with isolated consumers. This is supposed to optimize the efficiency of power generation and also deliver ac power at any desired voltage and frequency. Commonly, a diode-bridge rectifier with an electrolytic capacitor is used to convert the generator ac output into dc for the cascaded inverter. The capacitor is used to reduce ripples in the dc voltage. Such converters are simple and the least expensive. However, besides producing unregulated dc voltage, the rectifier draws non-sinusoidal input current (generally peaky in nature) from the generator. This creates problems such as distortion of the ac voltage waveform, reduction in generator output capability owing to poor form factor of the current, stress on the filter capacitor due to peak pulse currents, and increased losses in the diode bridge.

With only the capacitor on the dc side, it will be charged to the peak of the supply voltage and then start discharging into the load till the rectified voltage exceeds the capacitor voltage. The current will now flow from the ac source charging the capacitor and supplying the load. Owing to the short charging period, the rectifier output current will be discontinuous and the supply current will consist of current pulses. An inductor placed on the dc side between the rectifier and the capacitor will considerably improve the current waveform and reduce the output voltage ripple. The inductor prolongs the flow of output current, turning the input phase current into a quasi-square wave.

In order to attenuate the magnitude of the undesirable harmonic components being passed into the supply system, some tuned LC harmonic filters may be provided on the supply side. A practical problem arises when the supply frequency changes, as in the case of the variable-speed wind generator. Moreover, the filter components, particularly for the low-order harmonics, may become very large, both in size and weight, if the desired power factor approaches unity and the desired total harmonic distortion is to be low. To overcome the problems of harmonics and improve the power factor, active current wave shaping techniques have been developed to convert the harmonic-rich current into a close sinusoidal current in phase with the source voltage. An electronic circuit arrangement for reducing harmonics is known as the *active filter* or *line conditioner*. A power line active filter employs an additional power converter stage (PWM boost chopper) or a PWM converter in parallel with the non-linear load.

Single-phase rectifier

Figure 4.31(a) shows the approach to implement active wave shaping of the input line current waveform of a single-phase diode-bridge rectifier for unit power factor. The rectifier is followed by a PWM boost chopper. The control scheme ideally aims at unit power factor with sinusoidal current and a constant dc output voltage, which should be higher than the peak supply voltage. The actual dc output voltage is compared with a reference value, and the amplified voltage error is multiplied by a signal proportional to the absolute value of the instantaneous supply voltage to generate the commanded current wave for the inductor. The chopper switch is then controlled by a hysteresis comparator so as to keep the measured current within a tolerance band around the commanded sine wave current. The resulting i_s waveform is shown in Fig. 4.31(b).

The current mode control of the switch can also be carried out at a constant frequency. The switch is turned on by a fixed frequency clock at the beginning of the switching period and turned off when the measured current I_L equals the commanded current I_L^*. The current starts decreasing and the switch is again turned on at the commencement of the next cycle.

Three-phase rectifier

The basic circuit configuration to obtain near sinusoidal supply current at a high power factor for a three-phase diode bridge is shown in Fig. 4.32(a). It is a three-phase, single-switch-operated, boost-type chopper rectifier with the boost inductor on the supply side. The boost chopper is made to work under the discontinuous inductor current mode at a fixed frequency much higher than the supply frequency. When the boost switch is turned on, all the phases are shorted through the diodes.

For very high switching frequency compared with the system, the impressed voltages in the shorted phases can be considered

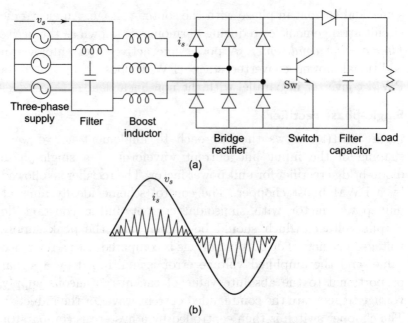

Fig. 4.32 (a) Schematic diagram for active-input, line-current wave shaping of a three-phase diode-bridge rectifier using a discontinuous inductor current mode boost chopper; (b) supply voltage and converter input current waveforms

almost constant within the switching period. In such a situation, the input phase currents will increase almost linearly. In order to keep the peak value of the input phase current very closely proportional to the instantaneous values of the respective phase voltages, the switch will be operated with constant on time. For this constant on time, within the time period of the supply, the envelopes of the peak currents will then be close to sinusoids in phase with the respective phase voltages, as shown in Fig. 4.32(b). High-frequency current harmonics, a result of the typical switching scheme, are easily filtered with a small input filter. The peak current values also depend on the boost switch duty cycle at constant frequency that is given by an output voltage controller. In the discontinuous current mode, as the current build-up starts from zero, sinusoidal variation of the peak input current implies that the average value of the bridge input current over the switching period also changes sinusoidally. As the scheme depends on the discontinuous inductor current mode, the switching frequency of Sw has to be selected such that the inductor current becomes zero before the next cycle.

Summary

The feasibility of the generation of electrical power by wind turbines is well established, and the aim is now to focus on the efficient operation of the system—both technically and economically. The key requirement nowadays is satisfied by power electronic circuits and systems. A power electronic interface is required to make the variable-voltage and/or variable-frequency output from a wind generator interactive with consumer and utility grids demanding quality supply, i.e., constant-voltage, constant-frequency supply with negligible harmonic content. This chapter introduces the power electronic devices and converter circuits related to the processing of wind electric power to make it compatible with the load requirements and the utility grid. After a brief introduction to the classification and components of power electronic converters, the salient features of various power electronic devices and their V-I characteristics are presented.

Rectifiers, inverters, and dc-to-dc and ac-to-ac converters are the basic requirements for processing the power generated from renewable energy sources. The advent of high-power semiconductor switches has made it possible for these converters to work efficiently and reliably under suitable control algorithms. In this chapter we introduce phase-controlled converters and ac voltage regulators, dc–dc PWM converters, and voltage-fed and current-fed inverters. Voltage-fed inverters include quasi-square wave, sine PWM, and current-regulated PWM inverters.

Each type of converter has their own merits and limitations. For example, a simple diode bridge may suffice to obtain dc output from a variable-voltage, variable-frequency permanent magnet ac generator, whereas a PWM rectifier may be required in another instance.

The principle of bidirectional operation (rectification and inversion) of a sine PWM converter is presented and its control strategy explained. Bridge diode input currents are rich in harmonics. Input current wave shaping methods for single-phase and three-phase diode bridges are given and discussed. This chapter has thus presented those aspects of power electronic converters which are considered relevant in wind electrical systems.

Problems

1. A three-phase diode bridge is supplied by a synchronous generator whose excitation emf is 1.06 p.u. and synchronous reactance is 0.25 p.u. Assuming continuous load current of 0.8 p.u., determine the percentage of the dc output voltage of its no-load voltage and the total rating of the rectifier. Neglect diode drops.

2. A three-phase sine PWM rectifier is to supply a 500-V, 10-kW dc load from a three-phase ac supply of 110 V, 50 Hz at unity power factor. The value of the ac supply line inductance is 5 mH. Calculate the amplitude modulation index of the sinusoidal reference and its phase angle in relation to the supply voltage. Assume ideal devices and a large capacitor at the rectifier output terminals for the load voltage to be ripple free. Neglect all losses.

3. A full converter comprising four thyristors supplies a resistive load of 20 Ω from a three-phase, 110-V, 50-Hz supply. The value of each ac line inductance is 5 mH. There is no filter capacitor at the output terminals. If the triggering angle is 30°, find (i) the rectifier output voltage and (ii) the fundamental active and reactive components of the supply current.

4. A single-phase boost PWM rectifier, as illustrated in Fig. 4.31, supplies 2200 W to a 220-V dc system from a single-phase, 110-V, 50-Hz supply. Determine the output capacitance and the value of the boost inductor to contain 220 V within ±2% and restrict the current ripple to 0.4 A. The switching frequency is 20 kHz and the power factor is close to unity.

5. A three-phase ac-to-dc converter, as shown in Fig. 4.32, consists of a boost inductor of value 40 µH in each phase, a

three-phase diode-bridge rectifier, a boost switch operating at a frequency of 25 kHz with a duty cycle of 0.5, and a filter capacitor of a large value. The converter supplies a resistive load of 100 Ω. Find the load voltage and the rms value of the fundamental component of the source current. Assume ripple-free voltage across the capacitor and lossless converter components. The input is the three-phase sinusoidal voltage of 110 V, 50 Hz.

5

Grid-connected and Self-excited Induction Generator Operation

There are two ways of exciting an induction generator. Based on the method of excitation, induction generators are classified into two basic categories, namely,

(a) constant-voltage, constant-frequency generators and
(b) variable-voltage, variable-frequency generators.

There are other ways of classifying induction generators, but these are generally related to the method of operation of the machine, based on certain control schemes.

In the constant-voltage, constant-frequency category, the generator derives its excitation from the utility bus as shown in Fig. 5.1(a). The generated power is fed to the supply system when the rotor is driven above synchronous speed. Machines with a cage-type rotor feed only through the stator and generally operate at low negative slip. But wound rotor machines can feed power through the stator as well as the rotor to the bus over a wide speed range.

Figure 5.1(b) presents the second type, which is analogous to a self-excited dc generator. A capacitor, when connected across the induction machine, helps build up the terminal voltage. But the

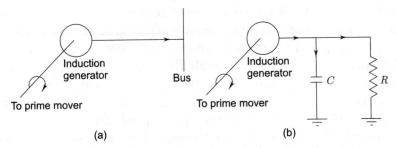

Fig. 5.1 Excitation systems: (a) line excitation, (b) self-excitation

building up of the voltage also depends on factors such as speed, capacitor value, and load. The squirrel cage machine is generally used as a self-excited induction generator.

5.1 Constant-voltage, Constant-frequency Generation

An induction machine in the generating mode operates fundamentally in the same manner as in the motoring mode except for the reversal of power flow. Consequently, the equivalent circuit and the associated performance equations, derived earlier using motoring conventions, are valid for all values of slip. With the stator winding remaining connected to the utility grid, if the rotor is driven by a prime mover above the synchronous speed in the direction of the air-gap field, the mechanical power of the prime mover is converted into electrical power.

5.1.1 Single-output System

Fixed-speed system

The system in the general sense implies the use of the squirrel cage induction generator, which provides the power output only through the stator winding. Figure 5.2 illustrates the configuration. It requires a grid-connected squirrel cage induction generator coupled to a turbine through a gear box. The gear steps up the rotor speed to a value matching a 50- or 60-Hz utility network. The generator always draws reactive power from the network. Capacitors are used to compensate this lagging VAR. These capacitors may cause the induction machine to self-excite, leading to over voltages at the time of the disconnection of the wind turbine from the electrical system if proper protective measures are not taken. Because of its coupling to the grid, the speed varies over a very small range above synchronous speed, usually around 1%. As the speed variation is small, the system is commonly known as a fixed-speed system. For such a system, the tip speed ratio λ varies over a wide range, making the rotor efficiency suffer at wind speeds other than the rated wind speed. The gear box ratio is selected for optimal C_p for the most frequent wind speed. In a well-designed system, fixed-speed operation can extract about 80% of the energy available from a fully variable speed system over a year. Fixed-speed wind turbines employing either blade pitch regulation or stall regulation to limit the power at high wind speeds are used. It is necessary to do so because if the input mechanical power is more than the power corresponding to the pull-out torque, the system becomes unstable.

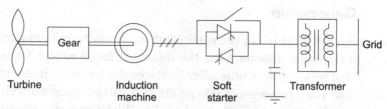

Fig. 5.2 Fixed-speed system with a squirrel cage induction generator

In a pitch-regulated system the electrical output power is regulated by a control system, which alters the blade pitch angle to extract the maximum energy at wind speeds below the rated wind speed; the power output is governed towards a limiting value at wind speeds above the rated speed. With stall regulation the blades are set at a constant pitch angle and the turbine enters the stall mode at high wind speed, thereby limiting the output power. Stall control is commonly applied in fixed-speed generators.

Appreciable generation at low wind speeds requires reduced rotor speed. To achieve this, one can use a two-speed cage-type induction generator with a stator winding arrangement for two different numbers of poles. The large number of poles is for low wind speed and the small number of poles is for high wind speed. An appropriately designed two-speed system can extract as high as 90% of the energy obtainable from a 100% variable-speed system over a year. With a two-speed system, the audible noise at lower wind speed is reduced.

Usually, the turbine accelerates the induction machine to synchronous speed using wind power; the machine is then connected to the grid. The direct connection of an induction machine to the supply produces high inrush current, which is undesirable, particularly in the case of electrical networks with low fault tolerance levels. Such a connection can also cause torque pulsations, leading to gear box damage. In order to reduce the magnetizing current surge, soft-start circuits utilizing phase-controlled antiparallel thyristors (ac voltage controllers), as illustrated in Fig. 5.2, are frequently employed to control the applied stator voltage when the induction machine is connected to the network. A few seconds later, when normal current is established, these starting devices are bypassed. Such ac voltage controllers can also be used for connecting the machine to the grid during acceleration from zero speed to the operating speed. This is particularly useful in stall-controlled systems.

Semi-variable-speed operation

The advantages of a grid-connected, fixed-speed squirrel cage generator are its lower capital cost, simple system configuration, and robust mechanical design. As the rotor speed is nearly constant, fluctuations in wind speed result in torque (power) excursions, which may lead to unwanted grid voltage fluctuation and strains on the turbine components. Wind gusts in particular lead to large torque variations.

Limited variable-speed operation in this single-output system can bring down the pulsations in grid power (voltage) and mechanical stress. If some of the generator shaft input (i.e., turbine output) can be dissipated in the rotor, the grid input power (which is essentially the power flow across the gap) can be levelled under fluctuating wind speed conditions. According to Eqns (3.11) and (3.40), the rotor electrical power is proportional to the slip. It then becomes possible to achieve speed control through control of the energy dissipated in a rotor resistor. Furthermore, the variation of rotor resistance with speed (i.e., with slip) in accordance with $(R_r + R_x)/s = $ constant [see Eqn (3.22)] keeps both the rotor current and the air-gap power (torque multiplied by the synchronous speed) constant. Hence the main aim of the control strategy will be to keep the rotor current at a set value,

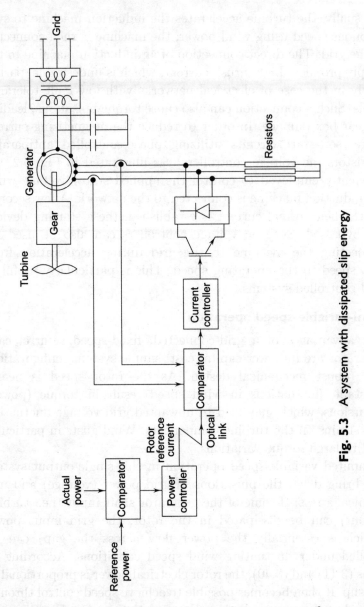

Fig. 5.3 A system with dissipated slip energy

irrespective of the speed variation within a range, for constant power output from the stator.

Figure 5.1.1 presents a configuration which includes a gear box, an induction motor with a three-phase stator winding, and a wound rotor with an electronically variable rotor resistance. The converter and the resistance rotate with the rotor. Control signals are sent to the rotating electronic parts by opto-electronic means. The rotor current reference comes from the comparison between the actual power and the reference power. The average resistance is varied between zero and the full value by continuously adjusting the duty cycle of the transistor switch. When the wind speed goes above the nominal value, the rotor current is held constant at a value corresponding to the rated power, by decreasing the duty cycle of the transistor switch. This will cause the generator speed to change, at the same time maintaining constant stator power. This configuration allows limited speed variation. The system shown in this figure is being marketed by Vestas of Denmark under the trade name OptiSlip with a maximum variation of 10% over nominal speed.

5.1.2 Double-output System with a Current Converter

With a slip-ring induction machine, power can be fed into the supply system over a wide speed range by appropriately controlling the rotor power from a variable-frequency source. The provision for bidirectional flow of power through the rotor circuit can be achieved by the use of a slip-ring induction motor with an ac/dc/ac converter connected between the slip-ring terminals and the utility grid. The basic configuration of the system is shown in Fig. 5.4. The system is known as a *double-output induction generator* (DOIG) because power can be tapped both from the stator and from the rotor. Figure 5.5 presents the main components of the solid-state system for the controlled flow of slip power at variable speed through current converters. The intermediate smoothing reactor is needed to maintain current continuity and reduce ripples in the link circuit. For the transfer of electrical power from the rotor circuit to the supply, converters I and II are operated, respectively, in the rectification and inversion modes. On the other hand, for power flow in the reverse direction, converter II acts as a rectifier and converter I as an inverter. The step-down transformer between converter II and the supply extends the control range of the firing delay angle α_2 of converter II.

The firing delay angle α_1 of converter I on the rotor side controls the phase difference between the injected rotor phase voltage and the rotor current, while the delay angle α_2 of converter II on the line side dictates the injected voltage into the rotor circuit.

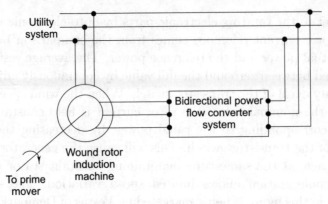

Fig. 5.4 Double-output induction generator system

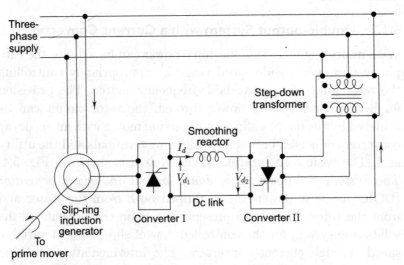

Fig. 5.5 Double-output system with direct current link

Line-commutated converters cannot generate leading VAR, and so, for maximization of the power output, α_1 should be set at 0° (rectification mode) in the supersynchronous region above the rated speed n_r (see Fig. 3.5) to draw power out of the rotor, and at 180° (inversion mode) in the subsynchronous region to inject power at the slip frequency into the rotor circuit. Power flow characteristics can be studied using ac as well as dc equivalent circuits. A reasonable estimate for the required variation in α_2 as a function of slip can be obtained by considering the dc voltage balance between the two sides of the smoothing reactor in the antiparallel bridge network.

To derive the equivalent circuits and analyse this system, we make the following assumptions.

(a) The magnetizing current and iron loss are neglected.
(b) The inverter-side transformer is assumed to be ideal.
(c) The commutation of the switching devices is assumed to be instantaneous and the device losses are neglected.
(d) The harmonic effects are ignored.

5.1.3 Equivalent Circuits

The ac equivalent circuit

Neglecting stator and rotor leakage impedance drops, the average voltage output of converter I at a slip s is given by

$$V_{d_1} = \frac{3}{\pi}\sqrt{6}V_2|s|\cos\alpha_1 \tag{5.1}$$

where V_2 is the slip-ring voltage (per phase) at standstill. For converter II, the average voltage output is given by

$$V_{d_2} = \frac{3}{\pi}\sqrt{6}\frac{V_1}{m_2}\cos\alpha_2 \tag{5.2}$$

m_2 being the turn ratio of the step-down transformer between converter II and the supply.

For dc voltage balance, neglecting the resistance drop in the dc-link smoothing inductor

$$V_{d_1} + V_{d_2} = 0 \tag{5.3}$$

This yields

$$\cos\alpha_2 = -\frac{m_2}{m_1}|s|\frac{V_2'}{V_1}\cos\alpha_1 \tag{5.4}$$

where V_2' is the stator-referred, slip-ring open-circuit voltage. Equation (5.4) can be used for the evaluation of α_2. Under the assumption of negligible stator impedance drops, V_2' equals V_1 and Eqn (5.4) becomes

$$\cos\alpha_2 = -\frac{m_2}{m_1}|s|\cos\alpha_1 \tag{5.5}$$

In fact, the presence of the dc-link circuit resistance demands

$$\cos \alpha_2 \geq -\frac{m_2}{m_1}|s| \cos \alpha_1 \tag{5.6}$$

Neglecting the losses in the semiconductor switches, the slip power, i.e., the rotor electrical power, is partly dissipated in the dc-link and rotor resistances, and the rest is fed back to the supply system through converter II. With reference to Fig. 5.5, the power input to converter II from the dc-link side (inverter operation) is

$$P_2 = -V_{d_2} I_d \tag{5.7}$$

Using Eqn (5.2) in Eqn (5.7), we get

$$P_2 = -\frac{3}{\pi} \sqrt{6} \frac{V_1}{m_2} I_d \cos \alpha_2 \tag{5.8}$$

For 120° conduction of each device in the converters, the fundamental rms component I_2 of the ac-side current of the converters (rotor or transformer secondary) and the dc-link current are related by

$$I_2 = \frac{\sqrt{6}}{\pi} I_d \tag{5.9}$$

From Eqns (5.8) and (5.9), the power fed back to the supply by the rotor is obtained as

$$P_2 = -3\frac{V_1}{m_2} I_2 \cos \alpha_2 \tag{5.10}$$

Referring the current I_2 to the stator side, we get

$$P_2 = 3\frac{m_1}{m_2} V_1 I_2' \cos \phi_2 \tag{5.11}$$

where ϕ_2 is the supplementary of the delay angle α_2 of converter II (i.e., $\phi_2 = \pi - \alpha_2$).

The rms value of the quasi-square wave rotor current is

$$I_r = \sqrt{\frac{2}{3}} I_d \tag{5.12}$$

The total secondary circuit copper loss

$$\begin{aligned} P_{Cu2} &= 3I_r^2 R_r + I_d^2 R_d \\ &= 3I_r^2 (R_r + 0.5R_d) \end{aligned} \tag{5.13}$$

The rotor rms current consists of the fundamental rms component and the higher harmonic rms components. Assuming that the torque is produced by the fundamental component of the rotor current, mechanical power can be expressed as

$$P_m = (\text{rotor-side electrical power with } I_2) \times \frac{1-s}{s}$$

$$= (P_{\text{Cu2}} \text{ due to } I_2 + P_2)\frac{1-s}{s}$$

$$= 3\left[I_2^2\left(R_r + 0.5R_d\right) + \frac{V_1}{m_2}I_2\cos\phi_2\right]\frac{1-s}{s} \qquad (5.14)$$

The air-gap power is

$$P_{\text{ag}} = P_2 + P_{\text{Cu2}} + P_m \qquad (5.15)$$

Substituting the expressions for P_2, P_{Cu2}, and P_m from Eqns (5.11), (5.13), and (5.14), using Eqns (5.9) and (5.12), and referring all the quantities to the stator side, we get

$$P_{\text{ag}} = 3\left[R_x' I_2'^2 + \frac{R_B'}{s}I_2'^2 + \frac{m_1}{m_2}\frac{V_1}{s}I_2'\cos\phi_2\right] \qquad (5.16)$$

where

$$R_x' = \left(\frac{\pi^2}{9} - 1\right)(R_r' + 0.5R_d')$$

$$R_B' = (R_r' + 0.5R_d')$$

Based on Eqn (5.16), we can draw a complete per-phase equivalent circuit, which includes the stator resistance and leakage reactance, as shown in Fig. 5.6. Now, the stator input power is given by

$$P_1 = 3I_1^2 R_s^2 + P_g \qquad (5.17)$$

With reference to Fig. 5.5 and following the motoring convention, the net electrical power output for the generating operation,

$$P_o = -(P_1 - P_2) \qquad (5.18)$$

The use of Eqns (5.11), (5.16), and (5.17) in Eqn (5.18) gives

$$P_o = -3\left[I_1^2 R_s + I_2'^2 R_x' + \frac{I_2'^2 R_B'}{s} + \frac{1-s}{s}\frac{m_1}{m_2}V_1 I_2'\cos\phi_2\right]$$

$$(5.19)$$

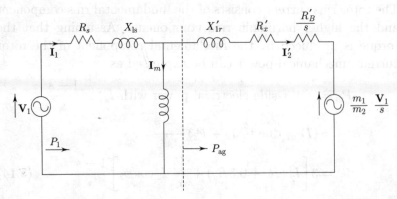

Fig. 5.6 Ac steady-state equivalent circuit per phase for the static Scherbius scheme

The dc equivalent circuit

In Fig. 5.7, the ac side (up to the rotor-side converter) represents the per-phase equivalent circuit of the induction machine referred to the rotor. The dc side of the equivalent circuit (to the right of the rotor-side converter) consists of the series resistance of the smoothing reactor and a voltage source representing the line-side converter.

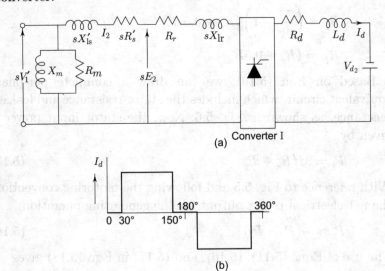

Fig. 5.7 (a) Rotor-referred induction motor equivalent circuit and dc link; (b) current waveform

It is convenient to refer the complete equivalent circuit to the dc side. First the ac-side resistances sR'_s and R_r are converted to their dc equivalents. To find out the dc equivalent resistance, recall that the input current of a phase-controlled converter has a quasi-square waveform as shown in Fig. 5.7(b). In this figure, the dc-side current is assumed to be ripple-free. The commutation overlap effect is also neglected. The rms value of the ac line current (in terms of the dc current) is given by

$$I_r = \sqrt{\frac{2}{3}} I_d \tag{5.20}$$

Now a resistance R_{dc} connected to the dc side of the phase-controlled converter I will be called the dc equivalent of the ac-side resistance if the power dissipated in R_{dc} equals the total power dissipated in all three phases. The ohmic power dissipated in all the three phases of the induction motor is

$$p_{Cu} = 3I_r^2 \left(|s|R'_s + R_r \right) \tag{5.21}$$

Using Eqn (5.20) in Eqn (5.21) gives

$$p_{Cu} = 2I_d^2 \left(|s|R'_s + R_r \right) \tag{5.22}$$

Therefore, the equivalent dc resistance of the induction motor when viewed from the dc-link side of converter I is

$$R_{dc} = 2|s|R'_s + 2R_r \tag{5.23}$$

Once the resistance portion of the ac side equivalent circuit is transferred to the dc side, the rest of the ac circuit along with the rotor-side converter (I) can be represented by a dc voltage source V_{d_1} in series with an equivalent internal resistance R_{d_1}. This takes into account the reduction in the mean output voltage of converter I caused by the induction motor reactance. The values of V_{d_1} and R_{d_1} in terms of the ac-side circuit parameters are given by

$$V_{d_1} = \frac{3\sqrt{6}}{m_1 \pi} |s| V_1 \cos \alpha_1 \tag{5.24}$$

and

$$R_{d_1} = \frac{3|s|}{\pi} \left(X'_{ls} + X_{lr} \right) \tag{5.25}$$

Being an inductive phenomenon, R_{d_1} does not represent a power loss component. Figure 5.8 shows the complete equivalent circuit referred to the dc side. Note that the entire slip power, i.e., the rotor-side electrical power, is supplied to the right of the dotted line. From the figure,

$$I_d = \frac{V_{d_1} + V_{d_2}}{R_d + R_{d_1} + 2(|s|R'_s + R_r)} \quad \text{for } I_d \geq 0 \qquad (5.26)$$

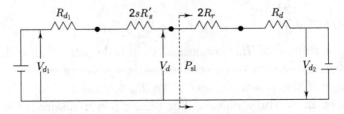

Fig. 5.8 The dc equivalent circuit referred to the rotor side

The condition $I_d \geq 0$ is required because the rotor-side and the line-side converters allow only unidirectional current. Substituting the expressions for V_{d_1}, V_{d_2}, and R_{d_1} in Eqn (5.26) and arranging the terms, we get

$$I_d = \frac{3\sqrt{6}V_1\,(m_2|s|\cos\alpha_1 + m_1\cos\alpha_2)}{m_1 m_2[\pi(R_d + 2R_r) + |s|\,(2\pi R'_s + 3X'_{ls} + 3X_{lr})]} \qquad (5.27)$$

Since $I_d \geq 0$,

$$m_2|s|\cos\alpha_1 + m_1\cos\alpha_2 \geq 0 \qquad (5.28)$$

The total electrical power in the rotor circuit, i.e., the slip power

$$P_{sl} = V_d I_d = (V_{d_1} - R_{d_1}I_d - 2|s|R'_s I_d)I_d \qquad (5.29)$$

i.e.,

$$P_{sl} = |s|\left\{ \frac{3\sqrt{6}V_1}{m_1\pi}\cos\alpha_1 - \left[\frac{3X'_{ls} + 3X_{lr}}{\pi} + 2R'_s\right]I_d \right\}I_d \qquad (5.30)$$

As the slip power P_{sl} is s times the air-gap power,

$$P_{ag} = \text{sign}(s)\left\{ \frac{3\sqrt{6}V_1}{m_1\pi}\cos\alpha_1 - \left[\frac{3X'_{ls} + 3X_{lr}}{\pi} \right.\right.$$
$$\left.\left. +2R'_s\right]I_d \right\}I_d \qquad (5.31)$$

The approximate rms value of the fundamental component of the stator current for level dc-link current I_d is

$$I_1 = \frac{\sqrt{6}}{m_1 \pi} I_d \tag{5.32}$$

Therefore, the stator input is

$$P_1 = P_{\text{ag}} + \frac{18 R_s}{m_1^2 \pi^2} I_d^2 \tag{5.33}$$

The power input to the utility grid through the supply-side converter (II) of the dc link is

$$P_2 = -V_{d_2} I_d$$

$$= -\frac{3}{\pi} \sqrt{6} \frac{V_1}{m_2} I_d \cos \alpha_2 \tag{5.34}$$

As the motoring convention has been followed, the net electrical power output for the generating operation,

$$P_o = -(P_1 - P_2)$$

$$= \left(P_{\text{ag}} + \frac{18 I_d^2}{m_1^2 \pi^2} R_s + \frac{3}{\pi} \sqrt{6} \frac{v_1}{m_2} I_d \cos \alpha_2 \right) \tag{5.35}$$

5.1.4 Reactive Power and Harmonics

The grid-connected induction generator draws its excitation from the power line to set up its rotating magnetic field for regeneration and thus always demands lagging reactive power. Such reactive power demand may adversely affect the network voltage level— particularly in weak public utility networks—and increase system losses. For large wind turbines driving induction generators, the voltage fluctuation and the flickering arising from power output variation may exceed the statutory limits of the utility system.

By applying the definition of reactive power $Q = V_1 I_1 \sin \phi_1$, to the T-circuit model of an induction motor [shown in Fig. 3.3(a)] the reactive power in one phase of the motor under the approximation $I_1 \approx I_2'$ becomes

$$Q = X_m^2 I_m^2 + I_2'^2 (X_{\text{ls}} + X_{\text{lr}}') \tag{5.36}$$

The use of Eqn (5.9) in Eqn (5.36) yields

$$Q = X_m^2 I_m^2 + \frac{6}{\pi^2} \frac{I_d^2}{m_1^2} (X_{\text{ls}} + X_{\text{lr}}') \tag{5.37}$$

The reactive power consumed by the electrical machine thus comprises a constant value, which is the magnetizing volt-ampere, and a parabolic value dependent on the dc-link current. A three-phase naturally commutated bridge converter on the rotor side is not capable of generating leading VAR. The lagging reactive power requirement is then transferred from the supply through the stator side of the machine, thus reducing the stator active power output for the same current loading. On the other hand, if the rotor-side converter is made a forced-commutated converter and its firing angle is made greater than 180° for operation below the rated speed and less than zero for operation above the rated speed, the reactive power demand of the machine can be met by the rotor-side converter. For controlled converter II, the phase angle between the fundamental alternating current and the ac sinusoidal current is equal to the firing delay angle α_2. As α_2 is always less than 180°, the fundamental lagging reactive power requirement of converter II comes from the electrical source through the step-down transformer, and varies with the operating point. Here too, if forced commutation is employed, unity or leading power-factor operation in order to improve the overall power factor of the system is possible. Whatever may be the firing strategy, the current-fed dc-link converter system requires an expensive choke and an extra commutation circuit for operation at synchronous speed (if it lies within the operating speed range). It also results in poor power factor at low-slip speeds. Besides, a current-fed dc link generates rectangular current waves, which inject low-order harmonics into the supply side of converter II, which are difficult to eliminate. The low-order harmonics from converter I injected into the rotor mmf produce variable-frequency stator current harmonics and torque harmonics.

For the reactive power requirement of the supply-side converter, various compensation schemes can be applied. A PWM current inverter (converter II) may be used to feed the mains with a quasi-sinusoidal current waveform with a phase shift ϕ, relative to the supply voltage, and can be controlled to generate reactive power. The reactive power control range of a current-fed dc-link inverter is, however, very limited.

5.1.5 Double-output System with a Voltage Source Inverter

The drawbacks of naturally commutated or line-commutated converters and low-frequency forced-commutated converters can be overcome by the use of dual PWM voltage-fed, current-regulated converters, connected back to back, in the rotor circuit, as shown in Fig. 5.9. PWM converters with dc voltage link circuits offer the following characteristics.

(a) Realization of the field-oriented control principle for decoupled control of the generator's active and reactive power.

(b) Low distortion in stator, rotor, and supply currents, owing to the shift of the harmonic spectra from lower to higher order, requiring a small-sized filter for attenuation of higher harmonics.

(c) Improvement in the overall system power factor through the control of the displacement factor between the voltage and current of the supply-side converter II.

(d) Operation at synchronous speed with direct current injected into the rotor from the dc voltage link circuit.

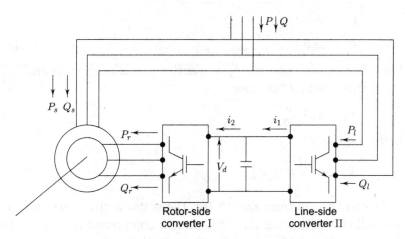

Fig. 5.9 Power flow in the slip power control scheme with dc voltage link

Decoupled control of the active and reactive power of an induction generator is carried out in a synchronously rotating d-q

axis frame with the d-axis of the frame overlapping the stator flux vector position. With such orientation of the d-axis,

$$\lambda_{qs}^e = 0 \tag{5.38}$$

Applying this condition to machine equations Eqn (3.49), (3.50), and (3.53), the defining equations of dynamical behaviour in such a reference frame can be written as

$$v_{ds}^e = R_s i_{ds}^e + p\lambda_{ds}^e \tag{5.39}$$

$$v_{qs}^e = R_s i_{qs}^e + \omega_e \lambda_{ds}^e \tag{5.40}$$

$$\lambda_{ds}^e = L_s i_{ds}^e + M i_{dr}^e \tag{5.41}$$

$$0 = L_s i_{qs}^e + M i_{qr}^e \tag{5.42}$$

Since the stator is connected to the grid and the stator resistance effect is small, the level of stator flux can be considered to be constant, decided by the magnitude and frequency of the grid voltage. Thus, for stator field orientation and $R_s \approx 0$,

$$\lambda_{ds}^e = \lambda_s \tag{5.43}$$

which is a constant. This reduces Eqns (5.39) and (5.40) to

$$v_{ds}^e = 0$$

$$v_{qs}^e = \omega_e \lambda_s \tag{5.44}$$

From Eqns (3.63) and (5.44), the active power at the terminals of the stator winding becomes

$$P_s = \omega_e \lambda_s i_{qs}^e \tag{5.45}$$

Using Eqn (5.42),

$$P_s = -\frac{M}{L_s} \omega_e \lambda_s i_{qr}^e \tag{5.46}$$

As can be seen from Eqn (5.46), the stator active power can be controlled by adjusting the rotor current component orthogonal to the stator flux. This subsequently controls the torque, as shown by Eqn (5.47) below.

For the assumed field orientation (5.38) and from Eqn (3.56), the electromagnetic torque becomes

$$T_e = \frac{P}{2} \lambda_s i_{qs}^e$$

which, using Eqn (5.42), can be expressed as

$$T_e = -\frac{P}{2}\frac{M}{L_s}\lambda_s i_{qr}^e \tag{5.47}$$

This expression for torque, when multiplied by the synchronous speed ω_s in mechanical radians per second, becomes the stator power expression (5.46).

Using Eqn (5.44) in Eqn (3.60), the reactive power at the stator winding terminals becomes

$$Q_s = \omega_e \lambda_s i_{ds}^e \tag{5.48}$$

which, using Eqn (5.41), becomes

$$Q_s = \frac{\omega_e \lambda_s}{L_s}(\lambda_s - M i_{dr}^e) \tag{5.49}$$

Therefore, the stator reactive power can be regulated by controlling the d-axis component of the rotor current i_{dr}^e. Thus the control of P_s and Q_s becomes essentially decoupled if i_{dr}^e and i_{qr}^e can be independently regulated.

The key to the implementation of the field-oriented approach is the determination of the stator flux angle θ_e. The machine terminal voltages and currents are sensed, the stationary two-axis components are evaluated using Eqn (3.48), and then Eqns (3.49) and (3.50) are applied with the condition $\omega_e = 0$ for the stationary reference frame to obtain the stator flux linkage space-vector $\vec{\lambda}_s$ in the manner outlined below:

$$\lambda_{ds}^s = \int \left(v_{ds}^s - R_s i_{ds}^s\right) dt$$

$$\lambda_{qs}^s = \int \left(v_{qs}^s - R_s i_{qs}^s\right) dt \tag{5.50}$$

$$\vec{\lambda}_s = \lambda_{ds}^s + j\lambda_{qs}^s$$

$$= \lambda_s \angle \theta_e$$

The total copper loss in the machine is

$$P_{\text{Cu}} = \left(i_{as}^2 + i_{bs}^2 + i_{cs}^2\right) R_s + \left(i_{ar}^2 + i_{br}^2 + i_{cr}^2\right) R_r \tag{5.51}$$

which, using Eqn (3.47), can be expressed as

$$P_{\text{Cu}} = \left(i_{ds}^{e2} + i_{qs}^{e2}\right) R_s + \left(i_{dr}^{e2} + i_{qr}^{e2}\right) R_r \tag{5.52}$$

The use of Eqns (5.41) and (5.42) in Eqn (5.52) gives

$$P_{\text{Cu}} = \left(R_s + \frac{L_s^2}{M^2} R_r \right)(i_{ds}^{e2} + i_{qs}^{e2}) + \frac{\lambda_s^2}{M^2} R_r - 2\frac{\lambda_s L_s R_r}{M^2} i_{ds}^e$$

(5.53)

Referring to Eqn (5.43), the term λ_s in Eqn (5.53) remains unchanged owing to constant supply voltage [vide Eqns (3.62) and (5.44)], while i_{qs}^e is governed by the stator active power (5.45) or the generator reaction torque. In wind energy conversion systems, a specific torque–speed relation is prescribed for capturing maximum power from the wind. Hence under the above constraint for minimum copper loss,

$$i_{ds}^e = \frac{L_s R_r \lambda_s}{R_s M^2 + R_r L_s^2}$$

(5.54)

i_{ds}^e controls the stator reactive power. Hence, from Eqn (5.48), the optimal stator reactive power based on the minimum copper loss criterion is

$$Q_s = \omega_e \frac{L_s R_r \lambda_s^2}{R_s M^2 + R_r L_s^2}$$

(5.55)

A vector control approach is used with the supply-side converter (i.e., converter II) for independent control of the active and reactive power between the supply and converter II. In the same stator-flux-oriented *d-q* reference frame, the following equations apply for the supply side of converter II:

$$P_l = v_{qs}^e i_{ql}^e + v_{ds}^e i_{dl}^e$$

$$Q_l = v_{qs}^e i_{dl}^e - v_{ds}^e i_{ql}^e$$

where i_{dl}^e and i_{ql}^e are the field-oriented currents on the line side of converter II. Referring to Eqns (5.44) and (3.62), $v_{ds}^e \approx 0$ and $v_{qs} = \sqrt{3}V$, and so

$$P_l = \sqrt{3}V i_{ql}^e$$

(5.56)

$$Q_l = \sqrt{3}V i_{dl}^e$$

(5.57)

Neglecting harmonic and converter losses, the following equations can be written (vide Fig. 5.9):

$$P_r = V_d i_2$$

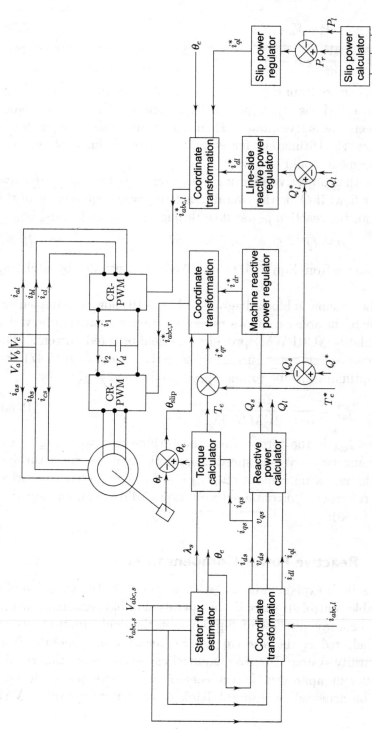

Fig. 5.10 Block diagram of the field-oriented control strategy

$$P_l = V_d i_1 \tag{5.58}$$

$$C\frac{dV_d}{dt} = i_1 - i_2$$

We can see from Eqns (5.57) and (5.58) that the dc-link voltage is controlled via i^e_{ql}, which in turn controls P_l. Any imbalance between the active powers P_l and P_r would raise the capacitor voltage V_d. Ultimately for stability of the dc-link voltage, the requirement would be $i_1 = i_2$.

Another objective of using converter II is to control reactive power flow. If Q^* is the overall reactive power requirement of the system, the reactive power flow through converter II must be

$$Q^*_l = Q^* - Q_s \tag{5.59}$$

We can see from Eqn (5.57) that Q^*_l can be controlled by adjusting i^e_{dl}.

The schematic block diagram in Fig. 5.10 shows the implementation of the above control strategy. Current-regulated pulse width modulation (CRPWM) provides the field-oriented currents. The reference q-axis rotor current i^{e*}_{qr} is derived from Eqn (5.47) and the optimum turbine torque–speed profile, and is given by

$$i^{e*}_{qr} = \frac{T_{\text{opt}}}{(-P/2)(M/L_s)\lambda_s} \tag{5.60}$$

where T_{opt} is the torque from the turbine corresponding to the optimum power versus speed characteristic of the given turbine. By the motoring convention, T_{opt} should be considered negative. The reference i^{e*}_{dr} is obtained from Eqn (5.49) in conjunction with Eqn (5.55).

5.2 Reactive Power Compensation

It has been explained in an earlier section how the adoption of a suitable control strategy in the operation of bidirectional converters to control the flow of slip power in the double-output system can help reduce the reactive power demand of the generator from the utility system. However, squirrel cage type generators require a different approach. Direct control of reactive power demand can be achieved by using a bank of capacitors, or other VAR compensators, located either centrally or near the wind farms. The network structure and the location of wind farms dictate the choice of the system. VAR compensators improve voltage stability, increase network capability, and reduce losses.

Various types of VAR compensators, with various features suitable for different applications, are in use.

The switched capacitor scheme

The switched capacitor scheme comprises a bank of parallel capacitors which are switched on and off by contactors in response to preset voltage levels. These are arranged in stages with a binary system for maximum flexibility of control. The speed of response is limited by the contactor closing time and is suitable for the discreet and slow control of the system voltage. For faster control, thyristor pairs are used to switch the capacitors (TSC) as shown in Fig. 5.11. Continuous control is not possible with the TSC scheme, as the capacitor would remain in the circuit for a full cycle before the thyristor switches off when the current reaches zero. In practice, current-limiting reactors are used in series with the capacitor banks to limit the current that may arise owing to the difference between the supply and capacitor voltages at the switching-on instant. In a three-phase system, capacitor banks are usually delta-connected.

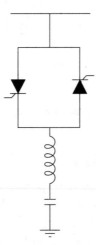

Fig. 5.11 VAR generation with a thyristor switched capacitor (TSC)

The thyristor controlled reactor

Continuous control of effective reactive power is possible if thyristor-phase-controlled reactors (TCRs), shown in Fig. 5.12, are used in parallel with fixed capacitor banks rated at the full-load reactive power demand of the induction generator. Variable VAR is realized by varying the firing angle between 90° and 180°. The excess reactive power from the capacitor bank at reduced load is absorbed by the reactor when the delay angle approaches 90°. In a three-phase system, the thyristor-controlled reactors are normally delta-connected to avoid triple harmonic components in the compensating line currents. Owing to the high cost of the system, it can be cost effectively employed at large wind farms rather than at each wind turbine site.

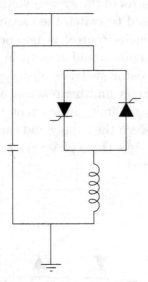

Fig. 5.12 A thyristor-controlled reactor (TCR) with a fixed capacitor

The static VAR compensator

The recent trend in reactive power control is based on the forced-commutated voltage source PWM converter. The basic circuit configuration is shown in Fig. 5.13(a). It is the static realization of the synchronous condenser. Inductors are included in series with the ac supply and a capacitor on the dc side. The dc capacitor is of considerably lower rating (typically 20%) and of small physical dimension compared to the ac capacitors that are used with conventional reactive power controllers. The main feature of this VAR compensator is that the converter can generate or absorb reactive power by controlling the switching pattern of the devices with gate turn-off capability, such as the GTO thyristor and IGBT. With a charged capacitor, the inverter produces a set of balanced voltages at the output terminal, which are controlled to be in phase with the corresponding ac system voltage, so that only reactive current can flow between the converter and the system. The basic principles of the control of reactive power flow are similar to those of the rotating synchronous condenser, and are shown in the phasor diagram of Fig. 5.13(b), for the current direction assumed in Fig. 5.13(a), for the fundamental component. When the inverter output voltage V_r is above the system voltage V_s, 90° lagging reactive current flows to the ac system and the converter therefore acts as a reactive power generator for the system. In other words, the converter appears as a capacitive load drawing leading current from the system. If the

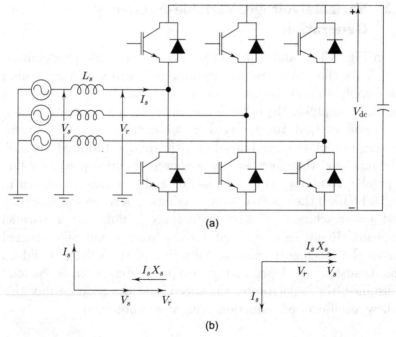

Fig. 5.13 Static VAR generator employing an inverter with a dc-side capacitor: (a) circuit, (b) phasor diagrams

inverter output voltage is decreased below that of the ac system, the reactive current will flow from the system to the converter, and thus the converter will appear as an inductive load on the system absorbing reactive power.

Since only the reactive power is involved, ideally the dc capacitor should retain its charge. The inverter simply helps connect different phases for the exchange of reactive currents between them. In a practical inverter, the devices are not lossless, and the inverter output voltage is made to lag behind the ac system voltage in case of leading VAR, and lead slightly for lagging VAR, so that the system can supply the small losses and the capacitor voltage can be maintained at the desired level. This principle can be used to increase or decrease the capacitor voltage, thereby controlling the inverter output voltage, for controlling the VAR generation or absorption. This static VAR generator can provide fast and continuous control of reactive power.

5.3 Variable-voltage, Variable-frequency Generation

From Fig. 3.3(b) and Eqn. (3.9), following the motoring convention, if the slip is negative, the motor input current will lag behind the supply voltage by an angle greater than $\pi/2$, as shown in Fig. 5.14, implying the induction machine to be a source of power. It is evident that the inverted motor current, i.e., the current flowing out of the machine, leads the motor terminal voltage. The machine, therefore, acts as a source of active power feeding a parallel combination of a capacitor and a resistor as shown in Fig. 5.14(b). If the reactive power oscillation between the capacitor and the machine's effective inductance, similar to a parallel resonant circuit, can be maintained, a voltage will be sustained across the machine terminals. Initiation of the voltage build-up and its sustenance depend on several parameters, such as the load resistance, the capacitance, the speed, and the residual flux this is how a self-excited induction generator is obtained.

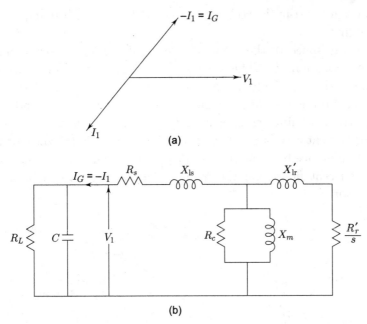

(a)

(b)

Fig. 5.14 Induction machine self-excitation: (a) phasor diagram, (b) basic equivalent circuit

5.3.1 The Self-excitation Process

For the self-excitation process to initiate, a capacitor bank of suitable size must be connected across the machine terminals, the core of which retains some residual flux. In order to understand the basic self-excitation process, let us refer to the circuit models shown in Figs 5.14(b) and 3.8. Combining them after neglecting the stator leakage impedance and the shunt resistance, a simplified circuit model of the self-excited induction generator under the no-load condition is obtained as shown in Fig. 5.15. The process

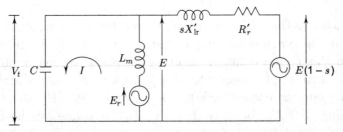

Fig. 5.15 Modified circuit model with speed emf in the rotor circuit

of voltage build-up is explained in the following with reference to this figure.

For any speed of the rotor, the residual flux generates a small synchronous emf E_r. The steady-state magnitude of the current through the $L_m C$ circuit is such that the difference between the synchronous saturation curve (voltage across L_m) and the capacitor load line, as shown in Fig. 5.16, at this value of the stator current equals E_r. At this stage, the slip s being zero for no speed difference between the rotor and the air-gap flux, no induced rotor current flows and the machine operates as a synchronous generator.

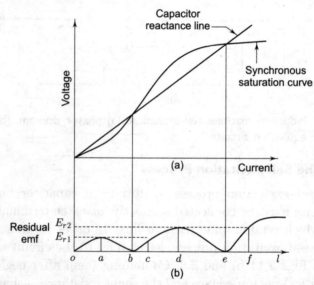

Fig. 5.16   Building up of voltage in a self-excited induction generator: (a) the capacitor load line and the saturation curve, (b) the difference between them

If E_r is less than E_{r1}, the machine operates in the stable steady state in the synchronous mode over the region oa. An increase in I in this region demands more synchronous voltage than the residual emf E_r. Consequently, the increased I is not sustained and the current comes back to its original value. By the same reasoning, if E_r is between E_{r1} and E_{r2}, a stable synchronous mode of operation is observed over the region cd. For $E_r \geq E_{r2}$, stable synchronous operation takes place from the point f onward. The regions ac and df are unstable, where, for the residual emf equal to E_{r1}, or E_{r2}, the machine terminal voltage rises owing to synchronous self-excitation, before entering the next stable region. In the stable regions, the machine operates as a self-excited synchronous generator.

The possibility of a changeover from synchronous generator operation to the self-excited asynchronous generator mode occurs in the region where the saturation curve emf is greater than the capacitor voltage. While the machine operates in the synchronous mode, any disturbance initiates an oscillation in the LC resonance circuit formed by the machine terminal capacitance and the magnetizing inductance at the natural angular frequency $\omega_n = 1/\sqrt{L_m C}$. Only at the points b and e does ω_n equal the synchronous frequency ω_1. Between the points b and e, the synchronous inductive reactance is greater than the capacitive reactance. Hence, the natural frequency ω_n of oscillation is lower than the rotational (i.e., synchronous) frequency ω_1. The air-gap flux associated with the oscillating current rotates at a speed lower than that of the rotor, implying a negative value of the slip. The corresponding rotational emf $E(1 - s)$, which exceeds E, drives a current into the stator circuit, building up the terminal voltage.

The machine now enters the asynchronous generating mode. An unstable oscillatory condition between the capacitor and the magnetizing reactance still persists owing to a continuous fall in the effective value of the magnetizing reactance as the terminal voltage rises. The natural frequency of oscillation progressively increases, and sustained oscillation is reached when the capacitive reactance is close to, but still less than, the magnetizing reactance near the point e. The small negative slip compensates the losses in the stator circuit. With a resistive load connected across the capacitor, the circuit must be underdamped to initiate the asynchronous generating mode.

5.3.2 Circuit Model for the Self-excited Induction Generator

In contrast to the source-connected generator, the capacitor self-excited induction generator presents a unique aspect—the main flux saturation assumes fundamental importance in establishing the equilibrium, that is, in determining the voltage level and the output frequency of the generator, for a given load, speed, and excitation capacitance. The capacitor is required to provide the reactive power, and no external constraint (frequency and/or flux) is imposed on the system. An uncontrolled self-excited induction generator shows considerable variation in its terminal voltage, degree of saturation, and output frequency under varying load conditions. For a systematic study of the behaviour of such a stand-alone induction generator with a variable-frequency output, it is convenient to base the analysis on a circuit model whose parameters are defined in terms of a base frequency.

Figure 5.17 shows the per-phase, steady-state, stator-referred equivalent circuit of a self-excited induction generator connected to a resistive–inductive load. A capacitor of capacitance C_0 is connected to provide the excitation VAR.

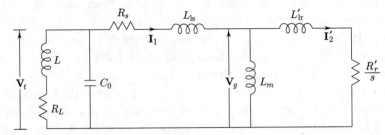

Fig. 5.17 Circuit model for a self-excited induction generator at stator frequency

With reference to Fig. 5.17, the following voltage equations in phasor notation can be written:

$$\mathbf{V}_t = R_s\mathbf{I}_1 + j\omega L_{ls}\mathbf{I}_1 + \mathbf{V}_g \tag{5.61}$$

which may be rewritten as

$$\mathbf{V}_t = R_s\mathbf{I}_1 + jFX_{ls}\mathbf{I}_1 + \mathbf{V}_g \tag{5.62}$$

where $F\,(=\omega/\omega_b)$ is the per-unit frequency and $X_{ls} = \omega_b L_{ls}$; w_b is the base angular frequency. Dividing Eqn (5.62) by F yields

$$\frac{\mathbf{V}_t}{F} = \frac{R_s}{F}\mathbf{I}_1 + jX_{ls}\mathbf{I}_1 + \frac{\mathbf{V}_g}{F} \tag{5.63}$$

The per-unit slip is

$$s = \frac{n - n_r}{n} \tag{5.64}$$

where n_r is the synchronous speed corresponding to the generated terminal frequency. The slip given in Eqn (5.64) may be expressed as

$$s = \frac{F - v}{F} \tag{5.65}$$

where v is the per-unit speed (n_r/n_b) defined with respect to the synchronous speed corresponding to the base frequency f_b and F is as defined earlier.

The voltage equation for the rotor circuit shown in Fig. 5.17 is

$$\mathbf{V}_g = \left(\frac{R'_r}{s}\right)\mathbf{I}'_2 + j\omega L'_{\mathrm{lr}}\mathbf{I}'_2 \tag{5.66}$$

In terms of the parameters F and v, Eqn (5.66) becomes

$$\frac{\mathbf{V}_g}{F} = \frac{R'_r}{F - v}\mathbf{I}'_2 + jX'_{\mathrm{lr}}\mathbf{I}'_2 \tag{5.67}$$

At the stator terminals, the following current balance equation holds:

$$-\mathbf{I}_1 = j\omega C_o \mathbf{V}_t + \frac{\mathbf{V}_t}{R_L + j\omega L}$$

In terms of the parameter F, this equation becomes

$$\mathbf{I}_1 = -\frac{\mathbf{V}_t}{F}\left[\frac{1}{(-jX_{\mathrm{cb}}/F^2)} + \frac{1}{R_L/F + jX_{\mathrm{Lb}}}\right] \tag{5.68}$$

where X_{cb} $(= 1/w_b C_o)$ and X_{Lb} $(= w_b L)$ are, respectively, the reactances of the excitation capacitor and the inductor at the base frequency.

Equations (5.63), (5.67), and (5.68) redefine the parameters and the node voltages of the equivalent circuit shown in Fig. 5.17, as indicated in Fig. 5.18. This stator-referred equivalent circuit, mapped in terms of the base frequency, is commonly used for predicting the performance of a self-excited induction generator. In general, the speed and the load are the given parameters. The frequency, the excitation capacitance, and the magnetizing reactance constitute the set of unknown variables even when a desired terminal voltage is required to be maintained.

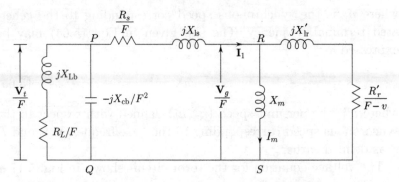

Fig. 5.18 The stator-referred circuit model of a self-excited induction generator normalized to the base frequency

5.3.3 Analysis of the Steady-state Operation

The equivalent circuit shown in Fig. 5.18 forms the basis for investigating the steady-state performance of a self-excited induction generator (SEIG) supplying a balanced load. There are two basic approaches, namely, the *loop impedance method* and the *nodal admittance method*, to defining the performance equations. The choice of the method is influenced by the objective of the analysis.

The loop impedance method

Since there is no emf source, applying Kirchhoff's emf law around the loop $SRPQ$ in the circuit of Fig. 5.18 yields

$$(Z_{QP} + Z_{PR} + Z_{RS})I_1 = 0 \tag{5.69}$$

For the stator current to exist under the self-excited state, the sum of the impedances must be zero, i.e.,

$$Z_{QP} + Z_{PR} + Z_{RS} = 0 \tag{5.70}$$

By equating the real and the imaginary parts independently to zero, we obtain two simultaneous non-linear equations:

$$f_1(F, X_{cb}, X_m) = 0$$
$$f_2(F, X_{cb}, X_m) = 0 \tag{5.71}$$

Two equations can yield only two unknown variables. The key unknown variable in determining the performance of an induction generator is the per-unit frequency F. The choice of the second unknown variable depends on the specific objective, i.e., the characterization of the problem. If the estimation of the minimum excitation capacitance for voltage build-up or the capacitance required for a specified value of the terminal voltage under a given load and speed is the requirement, X_{cb} is considered to be the unknown while X_m is given a specific value lying in the saturated region. The functions given by Eqns (5.71) then assume the forms

$$a_1 F^3 - a_2 F^2 - (a_3 X_{cb} + a_4)F + a_5 X_{cb} = 0 \qquad (5.72)$$

and

$$b_1 F^4 - b_2 F^3 - (b_3 X_{cb} + b_4)F^2 + (b_5 X_{cb} + b_6)F + X_{cb}b_7 = 0$$

$$(5.73)$$

where

$$a_1 = R_L TW + X_L T(R_s + R'_r)$$
$$a_2 = vT(R_L W + R_s X_L)$$
$$a_3 = R'_r(X_L + T) + T(R_s + R_L)$$
$$a_4 = R_L R_s R'_r$$
$$a_5 = vT(R_L + R_s)$$
$$b_1 = X_L TW \qquad\qquad (5.74)$$
$$b_2 = vb_1$$
$$b_3 = T(X_L + W)$$
$$b_4 = R_L T(R_s + R'_r) + R_s R'_r X_L$$
$$b_5 = vb_3$$
$$b_6 = vR_L R_s T$$
$$b_7 = R'_r(R_L + R_s)$$

in which

$$T = X_{ls} + X_m$$
$$= X'_{lr} + X_m$$
$$W = X'_{lr} + X_{ls}||X_m$$

Equations (5.72) and (5.73) can be solved numerically to obtain the values of X_{cb} and F. The initial value of C should be well above the value required for self-excitation under load and F should be less than the per-unit speed, though close to it.

A mathematical model similar to Eqns (5.72) and (5.73), in terms of the unknown variables X_m and F, can be formulated from Eqn (5.71) for given values of the excitation capacitance, the load, and the speed.

Once the values of X_{cb} (or X_m) and F are obtained, the following circuit relations can be used to determine the generator performance. The magnetization characteristic gives the values of the air-gap voltage V_g/F and the magnetizing current I_m corresponding to the chosen (evaluated) value of X_m.

$$\mathbf{I}_1 = \mathbf{I}_m + \frac{\mathbf{V}_g/F}{R'_r/(F-v) + jX'_{\mathrm{lr}}}$$

$$\mathbf{V}_t = \mathbf{V}_g + \mathbf{I}_1(R_s + jFX_{\mathrm{ls}})$$

$$\mathbf{I}_L = \frac{\mathbf{V}_t}{R_L + jFX_{\mathrm{Lb}}}$$

The output power is

$$P_o = 3|I_L|^2 R_L \tag{5.75}$$

Using the mathematical model presented above, the performance of an SEIG over a wide range of speeds, load impedances, and capacitances can be predicted from the equivalent circuit parameters and the magnetization characteristic at the base frequency. The consequences of the variation of the external parameters X_{cb}, v, and Z on the performance of the SEIG will be briefly presented in Section 5.3.4.

The nodal admittance method

The nodal admittance method has the advantage that it decouples the load and the excitation capacitor branch, and enables the per-unit frequency to be determined independent of the value of X_{cb}. Figure 5.18 can be redrawn as Fig. 5.19, where

$$R_{RS} = \frac{(F-v)R'_r X_m^2}{R_r^2 + (F-v)^2(X_m + X'_{\mathrm{lr}})^2}$$

$$X_{RS} = \frac{R'^2_r X_m + (F-v)^2 X_m X'_{\mathrm{lr}}(X_m + X'_{\mathrm{lr}})}{R_r^2 + (F-v)^2(X_m + X'_{\mathrm{lr}})^2}$$

The total impedance Z_{PS} of branch PRS is given by

$$Z_{PS} = \left(\frac{R_s}{F} + R_{RS}\right) + j(X_{ls} + X_{RS})$$

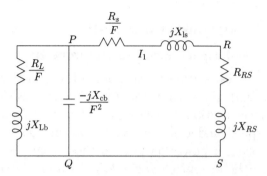

Fig. 5.19 The simplified equivalent circuit of a self-excited induction generator

By applying Kirchhoff's current law in Fig. 5.19, the sum of the currents at node P equals zero:

$$(Y_L + Y_C + Y_{PS})V_P = 0 \tag{5.76}$$

where Y_L, Y_C, and Y_{PS} are, repectively, the admittances of the series load circuit, the capacitive excitation circuit, and the machine equivalent circuit. Since V_P cannot be zero, for successful voltage build-up, it follows that

$$Y_L + Y_C + Y_{PS} = 0 \tag{5.77}$$

Equating the real and the imaginary parts independently to zero, the following equations are obtained:

$$\frac{R_L F}{R_L^2 + F^2 X_{Lb}^2} + \frac{F R_s + F^2 R_{RS}}{(R_s + F R_{RS})^2 + F^2(X_{ls} + X_{RS})^2} = 0 \tag{5.78}$$

$$\frac{F^2}{X_{cb}} - \frac{F^2 X_{Lb}}{R_L^2 + F^2 X_{Lb}^2} - \frac{F^2(X_{ls} + X_{RS})}{(R_s + F R_{RS})^2 + F^2(X_{ls} + X_{RS})^2} = 0 \tag{5.79}$$

Equations (5.78) and (5.79) are alternatives to Eqns (5.72) and (5.73), and are thus amenable to solution in the same manner. One marked difference can be noted: Eqn (5.78) is independent of X_{cb} and can be expressed as a 6th degree polynomial in F if the level of saturation, i.e., X_m, has been decided upon earlier. In certain studies, Eqns (5.78) and (5.79) are found to be more convenient.

5.3.4 The Steady-state Characteristics

Effect of external capacitance and load impedance

Figure 5.20 shows a typical variation of terminal voltage for resistive lead with the output power at a fixed speed for different values of the excitation capacitance. The curves suggest that, for a given speed and capacitance, an optimal load impedance exists for maximum power output. In these respects, the curves are similar to the output characteristics of a dc shunt generator with different field circuit resistances. The frequency decreases with the lead, but this variation is not significantly affected by the capacitance. Figure 5.21 exhibits the manner in which the capacitance requirement changes with the load and the power factor for constant terminal voltage at a fixed speed. The figure also indicates an increase in the VAR demand with decreasing load power factor.

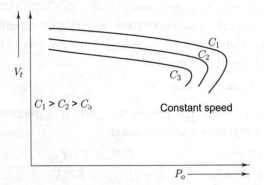

Fig. 5.20 Voltage regulation for different values of the excitation capacitance at constant speed

The group of curves shown in Fig. 5.22 shows the effect of controlling the excitation capacitance on the terminal voltage and the output power for a fixed load impedance at a constant speed. As is clear from these figures, there exist optimal capacitances that maximize the output power and terminal voltage for a given load impedance at a fixed speed.

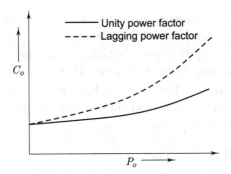

Fig. 5.21 Capacitance requirements for maintaining a constant voltage at the generator terminals for different power factors

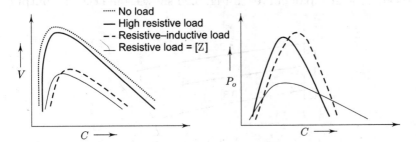

Fig. 5.22 Effect of the capacitance on the terminal voltage and the output power for a fixed load impedance

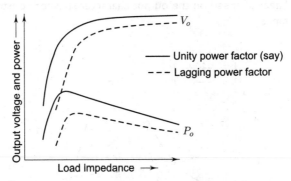

Fig. 5.23 Output voltage and power versus load impedance magnitude for different power factors at a fixed speed and constant capacitance

After the voltage has built up with a fixed value of the capacitance at a given speed, as the load impedance is gradually decreased, the output power initially increases from zero to a maximum value and then decreases till the terminal capacitance is unable to sustain self-excitation. Figure 5.23 shows the general trend in the variation of the output voltage and power for resistive and resistive–inductive loads.

Effect of speed

For power generation using wind energy, the speed of the prime mover varies over a wide range. For self-excitation, as indicated by the capacitive reactance value X_{cb}/F^2, the capacitor size is approximately proportional to the inverse of the square of the speed. The group of curves in Fig. 5.24 shows some typical output

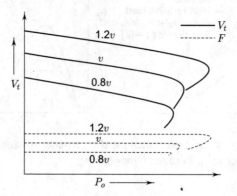

Fig. 5.24 Effect of speed on the output characteristics for constant capacitance

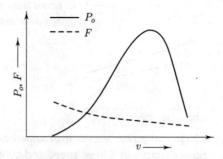

Fig. 5.25 Output power and frequency variation with speed

characteristics for different speeds under the constraint $X_{cb}/v^2 = $ constant. The terminal voltage and output frequency increase almost linearly with speed for the same power output over the working range. Figure 5.25 shows the output power versus speed curves for a given capacitance and load impedances. From the figure, it is clear that there exists a certain speed that maximizes the output power.

5.3.5 The Excitation Requirement

For successful excitation, the machine must operate in the saturated zone. The magnetizing reactance does not remain constant, but varies with the speed and the circuit parameters. It decreases with increasing saturation. Consequently, the minimum capacitance requirement for voltage build-up should correspond to the operating condition that results in a value of the magnetizing reactance $X_{s,max}$ slightly less than its unsaturated value X_{mu}. Hence, in this limit, the condition $X_m = X_{mu}$ yields the minimum value of the terminal capacitance below which the SEIG will fail to build up or sustain voltage, as the machine is gradually loaded following the build-up at a high saturation level.

To find the minimum capacitance required for self-excitation under the R-L load, X_{cb} is eliminated from Eqns (5.72) and (5.73). This gives a fourth degree polynomial in F:

$$p_4 F^4 + p_3 F^3 + p_2 F^2 + p_1 F + p_0 = 0 \tag{5.80}$$

Alternatively, Eqn (5.78) can be used. After a series of algebraic manipulations, Eqn (5.78) can be expressed as a sixth degree polynomial in F:

$$q_6 F^6 + q_5 F^5 + q_4 F^4 + q_3 F^3 + q_2 F^2 + q_1 F + q_0 = 0 \tag{5.81}$$

For minimum capacitance requirement, Eqn (5.80) or (5.81) is solved using $X_m = X_{mu}$ in the coefficients of F. The solution to Eqn (5.80) or (5.81), in general, yields two real roots and two pairs of complex conjugate roots. Only the real roots have any physical significance. A substitution of the real roots in Eqn (5.72) or (5.79) gives the corresponding set of capacitance values to make the induction generator operate at these frequencies. The minimum value of the capacitance corresponds to the larger value of F. The other value is impractical in the sense that the excitation

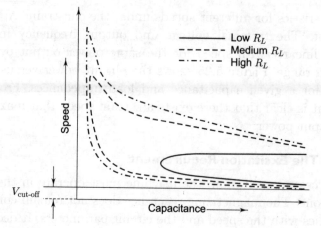

Fig. 5.26 Speed versus excitation capacitance for different loads

current goes well beyond the rated current of the machine. If Eqns (5.80) and (5.81) have no real roots, self-excitation is not possible. Figure 5.26 shows certain trends that can be obtained for a machine based on Eqn (5.80) or (5.81). For a given load there is a critical value of the capacitance below which self-excitation does not occur at any speed. This critical value decreases with increase in the load resistance. There is a minimum excitation speed as well as an upper limit for the load resistance beyond which self-excitation cannot be sustained with any value of the capacitance. Furthermore, there exists a cut-off speed below which the generator will never excite irrespective of any combination of the load and the capacitance. Similarly, for a given speed, there is a critical load below which excitation is not possible (Fig. 5.27).

To ascertain the capacitance requirement for self-excitation for a general R-L load, numerical solution is necessary. However, for the no-load case with the speed well above the critical speed, an acceptable simplified analytical model can be developed. Under the no-load condition, the terminal frequency is very close to the per-unit speed, making $R'_r/(F - v)$ quite large. The stator equivalent rotor current is very small and the machine current is predominantly reactive. Hence, referring to Fig. 5.18, for $R_L = \infty$ and $I_1 = I_m$,

$$\left(\frac{X_{\text{cb}}}{F^2} - X_{\text{ls}}\right) = X_m \tag{5.82}$$

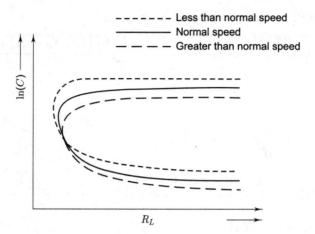

Fig. 5.27 Excitation capacitance versus load resistance at different speeds

i.e.,

$$X_{\text{cb}} = F^2(X_m + X_{\text{ls}}) \tag{5.83}$$

A very small stator copper loss is supplied by the generated power. Hence,

$$\frac{V_g^2(v - F)}{F^2 R_r'} = \frac{V_g^2}{F^2} \frac{R_s}{X_m^2 F} \tag{5.84}$$

That is,

$$F^2 - vF + \frac{R_s R_r'}{X_m^2} = 0 \tag{5.85}$$

Equations (5.83) and (5.85) yield

$$F_{\text{max}} = \frac{1}{2} \left(v + \sqrt{v^2 - \frac{4 R_s R_r'}{X_{\text{mu}}^2}} \right) \tag{5.86}$$

$$C_{\text{min}} = \left[\omega_b F_{\text{max}}^2 (X_{\text{mu}} + X_{\text{ls}}) \right]^{-1} \tag{5.87}$$

For a resistive load, it can be shown that the minimum per-unit speed v_{min} above which the induction generator will sustain excitation for a specific ratio of effective capacitance (at the base frequency) to the load resistance is given by

$$v_{\min} = \frac{4\left[\tau^2 + \tau\left(K_1 + K_2\right) + K_1 K_2\right]\left(X_r X_\sigma + X_m^2\right) X_\sigma X_r}{X_m^4}$$

(5.88)

where

$$\tau = \frac{X_{cb}}{R_L}$$

$$K_1 = \frac{\left[R_s + R_r\left(X_m/X_r\right)^2\right]}{X_\sigma}$$

$$K_2 = \frac{R_s X_r}{X_r X_\sigma + X_m^2}$$

The threshold per-unit frequency, corresponding to the condition in Eqn (5.88), is given by

$$F = \frac{v(2bX_r^3 + R_r X_m^2 X_r)}{2(bX_r^3 + R_r X_m^2 X_r)}$$

(5.89)

where

$$b = \tau x_\sigma + R_s$$

τ is not a direct physical variable. The same value of τ may be obtained with different combinations of X_{cb} and R_L. Hence the maximum value of τ does not bring out the critical values of X_{cb} or R_L. $\tau = 0$ indicates either $X_{cb} = 0$ or $R_L = \infty$. As $X_{cb} = 0$ indicates a short-circuit condition, $\tau = 0$ implies a no-load condition.

5.4   Effect of a Wind Generator on the Network

Many wind farms are connected to the local network at low, medium, or high voltage. The injection of wind power into the network has an impact on the voltage magnitude, its flicker, and its waveform at the *point of common coupling* (PCC).

The effect on the voltage magnitude depends on the 'strength' of the utility distribution network at the point of coupling as well as on the active and reactive power of the wind generator(s). The strength of the system at the point of coupling under consideration is decided by the short-circuit power, called the *fault level*, at that point. The short-circuit power is the product of the short-circuit current, following a three-phase fault at that point, and the voltage of the system. In fact, a power system comprises many interconnected power sources. The loads are fed through extended transmission and distribution networks. At the point of connection, as illustrated in Fig. 5.28(a), an equivalent ideal voltage source in series with an impedance Z_s may be assumed to replace the power system. Thus, the higher the fault current, the lower the source impedance. The wind farm with induction generators receives reactive power from the network and delivers real power to it. Without contribution from the wind generator, the fault level at the point of connection near the wind farm is

$$M = I_f V_s \qquad (5.90)$$

where

$$I_f = V_s / Z_s$$

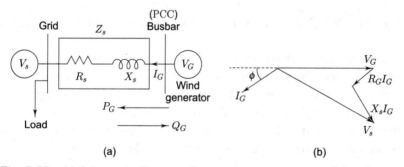

Fig. 5.28 (a) Schematic diagram of generator connection and distribution; (b) phasor diagram

Thus the fault level and hence the network strength are indicative of the source impedance. Areas of high wind velocity are suitable locations for wind farms. These areas are usually sparsely populated. Long transmission and distribution lines are normally required for connecting wind farms with the power system network. As a result, fault levels at the wind farms are generally low, making them weak electrical systems.

With reference to Fig. 5.28(b), if the phase difference between V_s and V_G is not large, the voltage at the PCC will be close to

$$V_G = V_s + R_s I_G \cos \phi - X_s I_G \sin \phi$$

$$= V_s + \frac{P_G R_s}{V_G} - \frac{Q_G X_s}{V_G} \tag{5.91}$$

Thus, at low power delivery, the voltage at the PCC reduces if the induction generator absorbs reactive power from the grid, while, at increased power flow, the voltage rises. *Flicker* is defined as the unsteadiness of the distribution network voltage. It may be caused by the continuous operation of a wind turbine or the switching operations of turbines. While operating, the rotor of a wind turbine experiences a cyclic torque variation at the frequency with which the blades move past the tower. This cyclic power variation may lead to flicker, and depends on the wind speed distribution at the site. While being connected to the network, the induction generator draws excessive current. Soft-start systems are usually employed to minimize the transient inrush current. However, at very high wind speeds, sudden disconnection of the wind generator from the distribution network may cause the voltage to dip, which cannot be avoided.

Interfacing variable-speed wind turbines with the network through electronic converters results in the injection of higher order harmonic currents, which distort the network voltage. This distortion is higher with weaker electric networks. It may be limited to a particular level, complying with the utility requirements, by installing harmonic filters or using PWM inverters.

Worked Examples

1. A three-phase, 440-V, four-pole, delta-connected, 50-Hz squirrel cage induction motor has the following star-equivalent circuit parameters: $R_s = 1.315$ Ω, $L_{ls} = 5.73$ mH, $R_r' = 1.1$ Ω, $L_{lr}' = 5.73$ mH. The magnetization characteristics at 1500 rpm are the following.

Induced emf (V)	Magnetizing current (A)
89.5	1
133	1.5
169.5	2
206	2.5
237	3
262	3.5
284.5	4
304	4.5
321.5	5
337.5	5.5
352	6
365	6.5

The machine is driven at 1000 rpm with a star-connected bank of 100-µF capacitors across its terminals.

(a) Will the machine self-excite? If it will, what will the open-circuit terminal voltage and frequency be?

(b) What is the minimum value of the capacitance required to cause self-excitation at this speed?

(c) What will be the minimum speed for the machine to self-excite with a 100-µF capacitor?

(d) With the machine remaining in the self-excited mode at 1000 rpm, what will be the minimum value of the load resistance with a 100-µF capacitor?

(e) What is the minimum value of load resistance that can be supported by the induction generator at 1000 rpm and the corresponding value of the excitation capacitor?

Solution

(a) The stator leakage reactance and the capacitive reactance at the base frequency 50 Hz are

$$X_{ls} = 2\pi \times 50 \times 5.73 \times 10^{-3} = 1.8 \ \Omega$$

$$X_{cb} = \frac{10^6}{2\pi \times 50 \times 100} = 31.83 \ \Omega$$

Under the no-load condition, the per-unit frequency F will be very close to the per-unit speed v, which equals $1000/1500 = 2/3$. Using $F = v$ gives

$$\frac{X_{cb}}{F^2} - X_{ls} = 69.81 \ \Omega.$$

A load line corresponding to 69.81 Ω intersects the saturation curve at 290 V. The machine is thus capable of self-excitation. Therefore,

$$\frac{E}{F} = 290$$

The induced emf per phase at 1000 rpm is

$$E = \frac{2}{3} \times 290 = 193.3 \text{ V}$$

The terminal voltage V_t will be the magnetizing current multiplied by the capacitive reactance at the operating frequency. The magnetizing current, as noted from the saturation characteristic corresponding to 290 V, is 4.15 A.

$$\frac{V_t}{F} = \frac{X_{cb}}{F^2} I_m$$

i.e.,

$$V_t = 31.83 \times \frac{3}{2} \times 4.15$$

$$= 198 \text{ V}$$

The magnetizing reactance at the existing level of saturation is 69.8 Ω. Using Eqn (5.86), the per-unit, no-load terminal frequency

$$F = \frac{1}{2} \left[\frac{2}{3} + \sqrt{\left(\frac{2}{3}\right)^2 - \frac{4 \times 1.315 \times 1.1}{69.8^2}} \right]$$

$$= 0.6662$$

Therefore, the frequency is

$$f = 50 \times 0.6662 = 33.31 \text{ Hz}$$

(b) From the saturation characteristic, the limiting value of the magnetizing reactance is taken to be 85 Ω. From Eqns (5.86) and

(5.87), respectively,

$$F_{max} = \frac{1}{2}\left[\frac{2}{3} + \sqrt{\left(\frac{2}{3}\right)^2 - \frac{4 \times 1.315 \times 1.1}{85^2}}\right]$$

$$= 0.6664$$

and

$$C_{min} = \left[2\pi \times 50 \times 0.6664^2(85 + 1.8)\right]^{-1}$$

$$= 82.5 \ \mu F$$

(c) At the minimum speed, 100 μF must be the minimum capacitance. Therefore, from Eqn (5.83),

$$F_{max}^2 = \sqrt{\frac{31.83}{85 + 1.8}}$$

$$= 0.605$$

Solving Eqn (5.85) for v gives

$$v = \frac{(1.315 \times 1.1/85^2) + 0.605^2}{0.605}$$

$$= 0.6053$$

This will be the minimum per-unit speed at which the machine can self-excite with a 100 μF capacitor. So $v_{min} = 0.6053 \times 1500 = 908$ rpm.

(d) We can solve this part of the problem by the nodal admittance method discussed in Section 5.3.3. Figure 5.29 (cf Fig. 5.18) shows the equivalent circuit with the given parameters and the unknown resistive load R_L at a per-unit frequency F.

Figure 5.30 shows the plots of the reciprocal of the imaginary part of the admittance of branch PS and $31.84/F^2$ against the per-unit frequency F. At $F = 0.649$, the inductive admittance of branch PS and the capacitive admittance become equal in magnitude. Power balance demands that

$$\frac{R_L}{F} = -\frac{1}{\text{Re}(Y_{PS})}$$

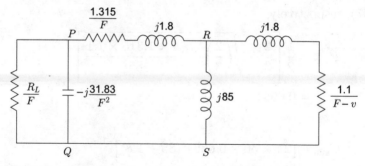

Fig. 5.29 Equivalent circuit of the self-excited induction generator

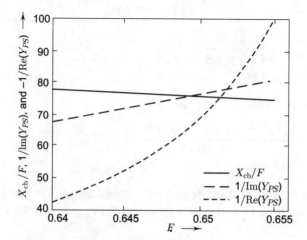

Fig. 5.30 Variation of the capacitive and the inductive admittances with frequency

where 'Re' denotes the real part. At $F = 0.649$,

$$\frac{R_L}{F} = 64.0$$

Therefore,

$$R_L = 64 \times 0.649$$

$$= 41.5 \ \Omega$$

(e) Substituting the values of the machine parameters in Eqn (5.88) and taking $v_{\min} = 1000/1500$ gives the following quadratic equation in τ:

$$\tau^2 + 0.683\tau - 2.48 = 0$$

which yields $\tau = 1.27$. From Eqn (5.89), the per-unit frequency at this limiting value of τ is

$$F = \frac{2}{3} \times \frac{7.2852}{2 \times 3.9868} = 0.609$$

Applying the condition stated in Eqn (5.70) to the circuit in Fig. 3.7(b), which is the inverse-Γ circuit model of the induction generator with a parallel RC circuit at the stator terminals, the following equation is obtained by equating the real parts of Eqn (5.70) to zero:

$$\frac{R_s}{F} + \frac{R'_r X_m^2 (F - v)}{R'^2_r + (F - v)^2 X'^2_r} + \frac{\tau X_{cb}}{F(F^2 + \tau^2)} = 0.$$

Substituting the values of the machine parameters, the per-unit frequency F, τ, and v, we get

$$2.16 - 17.45 + 1.05 X_{cb} = 0$$

$$X_{cb} = 14.55 \ \omega$$

Therefore,

$$R_L = \frac{X_{cb}}{\tau}$$

$$= \frac{14.55}{1.27} = 11.45 \ \Omega$$

2. A three-phase, 415-V, star-connected, four-pole, 50-Hz, slip-ring induction motor runs at a speed of 1447 rpm when operating at its rated load. The motor equivalent circuit parameters referred to the stator are $R_s = 0.075 \ \Omega$, $R'_r = 0.063 \ \Omega$, $R_m = 68 \ \Omega$, $X_m = 6.3 \ \Omega$, $X_{ls} = 0.305 \ \Omega$, and $X'_{lr} = 0.305 \ \Omega$. The stator-to-rotor turn ratio is 1.2. The machine is driven at supersynchronous speed and the power flow into the 415-V utility system is controlled by the static Scherbius scheme shown in Fig. 5.5, where a full-wave diode-bridge rectifier constitutes converter I. Converter II is connected to the source through a transformer with a turn ratio of 1.2. The resistance of the smoothing reactor of the dc link is 0.15 Ω.

(a) With the slip rings short circuited, how much power can be fed to the system for the same rated current and what should be the speed of the induction generator? Determine the efficiency of the generator.

(b) The induction machine is driven at 1800 rpm. Calculate the output power and the efficiency when harmonic currents and the dc-link resistance are (i) neglected and (ii) considered.

Solution

(a) In order to simplify calculations and still be able to predict performance accurately, we use the approximate equivalent circuit with an external rotor resistance R_x as shown in Fig. 5.31. For shorted slip rings, R_x is zero.

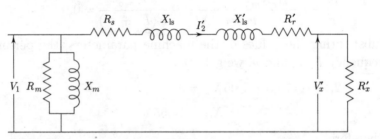

Fig. 5.31 Equivalent circuit with diode bridge rectifier at slip rings

Let s_g be the required slip for the generating operation and s_m that for the motoring operation. Then,

$$s_m = \frac{1500 - 1447}{1500} = 0.0353$$

For the same current values,

$$\left(R_s + \frac{R_r'}{s_g}\right)^2 = \left(R_s + \frac{R_r'}{0.0353}\right)^2$$

Substituting the parameters gives

$$\left(0.075 + \frac{0.063}{s_g}\right)^2 = \left(0.075 + \frac{0.063}{0.0353}\right)^2$$

The solution to the above equation yields

$$s_g = -0.0326, 0.0353$$

The negative value of s_g is accepted for the generating operation. The stator equivalent rotor current is

$$I'_2 = \frac{415/\sqrt{3}}{(0.075 - 0.063/0.0326) + j(0.305 + 0.305)}$$

$$= -116.5 - j38.2$$

$$= 122.5 \angle -161.8°$$

The air-gap power is

$$P_{ag} = \frac{3 \times 122.5^2 \times 0.063}{-0.0326}$$

$$= -87.1 \text{ kW}$$

The stator copper and iron losses are

$$P_{Cu1} + P_{iron} = 3 \times 122.5^2 \times 0.075 + \frac{415^2}{68}$$

$$= 5.9 \text{ kW}$$

The generator power output

$$P_o = (P_{ag} + P_{Cu1} + P_{iron})$$

$$= -87.1 + 5.9 = -81.2 \text{ kW}$$

The negative sign appears because of the motoring convention. Therefore, the input to the system is 81.2 kW.

$$\text{Mechanical power input} = \frac{3 \times 122.5^2 \times 0.063}{0.036}(1 + 0.0326)$$

$$= 89.84 \text{ kW}$$

Efficiency,

$$\eta_G = \frac{81.2}{89.84} = 90.4\%$$

(b) (i) The effects due to current harmonics and the dc-link resistance are neglected.

As the slip rings are connected to a diode bridge, the fundamental component of the rotor current and the slip-ring voltage will be in phase. Thus, as far as the slip-ring terminals are concerned, the diode-bridge rectifier, i.e., converter I, has the same effect as the controlled resistors across the slip rings. Let this external resistance be R'_x per phase in the equivalent circuit shown in Fig. 5.31.

The slip at 1800 rpm is

$$s_x = \frac{1500 - 1800}{1500} = -0.2$$

For the same rotor current at two slips,

$$\frac{R'_r}{s_g} = \frac{R'_r + R'_x}{s_x}$$

Inserting the respective values gives

$$\frac{0.063}{-0.0326} = \frac{0.063 + R'_x}{-0.2}$$

or

$$R'_x = 0.3235 \ \Omega$$

The stator-referred rotor current is

$$I'_2 = \frac{415/\sqrt{3}}{[0.075 - (0.063 + 0.3235)/0.2] + j0.610}$$

$$= -116.3 - j38.2 = 122.5 \angle -161.8° \ \text{A}$$

As expected, we get the same rotor current as in part (a). The air-gap power is

$$P_{ag} = \frac{3 \times 122.5^2 \times (0.063 + 0.3235)}{-0.2}$$

$$= -87.05 \ \text{kW}$$

The actual slip-ring voltage per phase, i.e., the diode input voltage

$$V_x = \frac{I'_2 R'_x}{m_1}$$

The output voltage of converter I is

$$V_{d_1} = \frac{3\sqrt{6}}{\pi} \frac{I'_2 R'_x}{m_1}$$

The dc-link voltage from the inverter side is

$$V_{d_2} = \frac{3\sqrt{6}}{\pi} \frac{V_1 \cos \alpha_2}{m_2}$$

As $V_{d_1} + V_{d_2} = 0$,

$$\cos \alpha_2 = -\frac{m_2}{m_1} \frac{I_2' R_x'}{V_1}$$

$$= -\frac{1.2}{1.2} \frac{122.5 \times 0.3235}{415/\sqrt{3}} = -0.1654$$

Therefore, the inverter firing angle is

$$\alpha_2 = 99.5°$$

The total loss in the reflected external resistances will be the estimate for the output, P_{ring}, through the slip rings.

$$P_{\text{ring}} = 3 \times 122.5^2 \times 0.3235$$

$$= 14.56$$

The power input to the supply system, i.e., the generator output

$$P_o = -(P_{\text{ag}} + P_{\text{Cu1}}) + P_{\text{ring}}$$

$$= -(-87.05 + 5.9) + 14.56$$

$$= 95.71 \, \text{kW}$$

The mechanical power input is

$$P_m = \frac{3 \times 122.5^2 \times (0.063 + 0.3235)}{0.2}(1 + 0.2)$$

$$= 104.4 \, \text{kW}$$

Therefore, the generator efficiency is

$$\eta_G = \frac{95.71}{104.4} \times 100 = 91.6\%$$

(ii) The current harmonics and the dc-link resistance are considered.

It is assumed that the dc-link current is ripple-free. The rotor currents will then be quasi-square waves. For the same heating effect, the rms value of the quasi-square wave current should be equal to that of the specified sine wave current

The actual rated rotor current (fundamental)

$$I_2 = 122.5 \times 1.2 = 147 \, \text{A}$$

The stator per-phase voltage

$$V_1 = 415/\sqrt{3} = 239.6 \text{ V}$$

The rotor-referred parameters are $R_r = 0.063/1.2^2 = 0.0438 \ \Omega$, $R'_s = 0.075/1.2^2 = 0.0521 \ \Omega$, and $X'_{lr} = X'_{ls} = 0.305/1.2^2 = 0.212 \ \Omega$.

For the same heating effect, the equivalent dc-link current will be

$$I_d = \sqrt{\frac{3}{2}} I_2$$

$$= \sqrt{\frac{3}{2}} \times 147 = 180 \text{ A}$$

Substituting the numerical values of the parameters in rotor terms in Eqn (5.27), we get

$$180 = \frac{3\sqrt{6} \times 239.6(1.2 \times 0.2 + 1.2 \cos \alpha_2)}{1.2^2 \left[\pi(0.15 + 2 \times 0.0438) + 0.2(2\pi \times 0.0521 + 3 \times 0.212 + 3 \times 0.212)\right]}$$

The solution to the above equation for α_2 gives

$$\alpha_2 = 93.96°$$

From Eqn (5.31) for air-gap power,

$$P_{ag} = -\left[\frac{3\sqrt{6} \times 415/\sqrt{3}}{1.2 \times \pi} - \left(\frac{3 \times 0.212 + 3 \times 0.212}{\pi}\right.\right.$$

$$\left.\left. +2 \times 0.0521\right) \times 180\right] \times 180$$

$$= -67.57 \text{ kW}$$

From Eqn (5.35), the output power is

$$P_o = -67.57 \times 10^3 + \frac{18 \times 180^2}{\pi^2} 0.0521$$

$$+ \frac{3\sqrt{6}}{\pi} \frac{239.6}{1.2} 180 \cos 93.96$$

$$= 70.31 \text{ kW}$$

From Eqn (5.14), the mechanical input power

$$P_m = 3[147^2(0.0438 + 0.5 \times 0.15) + (239.6/1.2)$$

$$\times 147 \cos(180° - 93.96°)](1.0 + 0.2)/(-0.2)$$

$$= -82.76 \text{ kW}$$

The negative sign appears because of the motoring convention. Therefore, the efficiency of the system is

$$\eta_G = (70.31/82.76) \times 100 = 84.91\%$$

3. A wind energy conversion system feeds power to a three-phase, 415-V, 50-Hz utility system. The drive train comprises a fixed-pitch wind turbine, a step-up gear box, and a squirrel cage induction generator. The following data are available for the wind turbine and the squirrel cage machine.

Wind turbine

Rated power	30 kW
Rated wind speed	12 m s^{-1}
Cut-in wind speed	3.5 m s^{-1}
Cut-out speed	90 rpm
Turbine blade diameter	10.5 m
Gear box ratio	9

The power coefficient curve, C_p versus β, is shown in Fig. 5.32.

Induction generator

Rated power	22 kW
Rated voltage	415 V
Rated current	43 A
Frequency	50 Hz
Poles	8
Rated speed	720 rpm
Stator resistance	0.145 Ω
Stator leakage reactance	0.702 Ω
Rotor resistance	0.263 Ω
Rotor leakage reactance	0.702 Ω
Magnetizing reactance	15.9 Ω
Core loss	190.6 W

Calculate the power fed into the system as the wind speed varies from 5 to 10 m/s. Assume the air density ρ to be equal to 1.225 kg m^{-3}.

Solution

Stable operation occurs when the turbine torque T_W matches the generator reaction torque T_G transmitted through the gear box.

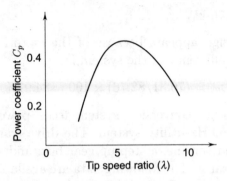

Fig. 5.32   Power coefficient versus the tip speed ratio curve

For a step-up gear box of ratio G,

$$T_W = GT_G \tag{5.92}$$

The turbine shaft torque is given by

$$T_W = \frac{1}{2}\rho\pi R^3 C_T V_w^2 \tag{5.93}$$

where the torque coefficient is

$$C_T = \frac{\text{power coefficient}}{\text{tip speed ratio}} = \frac{C_p}{\lambda} \tag{5.94}$$

in which

$$\lambda = \frac{\omega_t R}{V_w} \tag{5.95}$$

for the turbine speed ω_t (rad/s). The variation of C_T versus λ obtained from the plot of C_p versus λ in conjunction with Eqn (5.94) is shown in Fig. 5.33. The normal operating region is to the right of the maximum torque point $(C_{T,\max}, \lambda_{T,\max})$.

The variation of C_T beyond the point $\lambda_{T,\max}$ may be approximated by

$$C_T = C_1 - C_2\lambda \tag{5.96}$$

where, using the least-square curve fitting technique, C_1 and C_2 in Eqn (5.96) are obtained as $C_1 = 0.1618$ and $C_2 = 0.0138$.

From Eqn (3.21), the generator torque at a speed ω_r, neglecting the shunt branch, is

$$T_G = \frac{415^2}{(0.145 - 0.263/|s|)^2 + 1.404^2} \frac{0.263}{25\pi|s|} \tag{5.97}$$

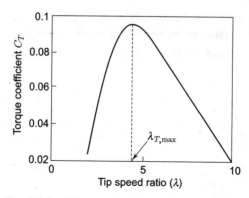

Fig. 5.33 Torque coefficient versus the TSR

where

$$|s| = \frac{\omega_r - \omega_s}{\omega_s} \qquad (5.98)$$

in which $\omega_s = 2\pi f/(P/2) = 25\pi$.

The use of Eqns (5.93), (5.95), (5.96), and (5.98) in Eqn (5.92) yields the following quadratic equation in V_w:

$$aV_w^2 + bV_w + c = 0 \qquad (5.99)$$

where

$$a = \frac{1}{2}C_1\rho\pi R^3/G = \frac{1}{2} \times 0.1618 \times 1.225 \times \pi \times 5.25^3/9 = 5.0058$$

$$b = -\frac{1}{2}C_2\rho\pi R^4\omega_r/G^2 = -\frac{1}{2} \times 0.0138 \times 1.225 \times \pi \times 5.25^4\omega_r/9^2$$

$$= 0.2491\omega_r$$

$$c = T_G \quad [\text{as given by Eqn (5.97)}]$$

For an assumed value of the generator speed ω_r, the wind speed V_w is obtained from Eqn (5.99). For the same value of ω_r, the power output from the stator, i.e., the power flow into the system, is given by

$$P_o = T_G\omega_s - 3I_s^2R_s - P_{\text{iron}}$$

where

$$I_s^2 = \frac{415^2}{3[(0.145 - 0.263/|s|)^2 + 1.404]^2}$$

Figure 5.34 shows the variation of P_o as a function of the wind speed V_w. For a change in the wind speed from 5 to 10 m/s, the power input to the system changes from 2.5 to 24.8 kW.

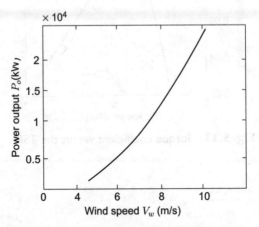

Fig. 5.34 Power output versus wind speed

4. A three-phase, 400-V, 50-Hz, four-pole wound rotor induction motor with a star-connected stator winding runs at a slip of 3.5% at the rated shaft torque with a short-circuited rotor. It has the following per-phase equivalent circuit parameters referred to the stator: $R'_r = 0.06\ \Omega$, $X'_{lr} = 0.12\ \Omega$, and $X_m = 6.12\ \Omega$. The stator-to-rotor turn ratio $m_1 = 2$. Stator impedance can be neglected. The machine is used in a static Scherbius drive configuration to extract power from a wind turbine. The rms rotor current reaches its rated value at $s = -0.5$.

(a) Calculate the shaft torque under this condition. The ratio of the gear between the wind turbine and the induction generator is adjusted such that at $s = -0.5$ the generator extracts maximum power at the rated wind speed.

(b) With the gear ratio adjusted as above, what is the minimum wind speed up to which maximum power can be extracted? Assume that up to this wind speed the generator rotor remains short-circuited.

Solution

The single-line diagram of the generating system, referred to the rotor, is shown in Fig. 5.35 (see also Fig. 5.5).

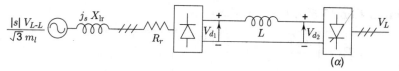

Fig. 5.35 The static Scherbius system

The dc equivalent circuit of the system with the stator impedance and dc-link resistance neglected is shown in Fig. 5.36.

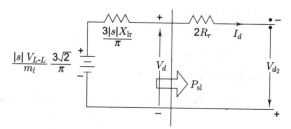

Fig. 5.36 The dc equivalent circuit

The per-phase supply voltage is

$$V_1 = \frac{400}{\sqrt{3}} = 230 \text{ V}$$

From the given data, the stator-referred rotor current [vide Eqn (3.8)] is obtained as

$$I'_{2,\text{rated}} = \frac{230}{\sqrt{0.06^2/0.035^2 + 0.12^2}}$$

$$= 134.4 \text{ A}$$

Therefore, the actual per-phase rotor current is

$$I_{2,\text{rated}} = 2 \times I'_{2,\text{rated}} = 268.8 \text{ A}$$

With a 120° quasi-square wave rotor current waveform, the per-phase rotor rms current and the dc-link current I_d are related as

$$I_2 = \sqrt{\frac{2}{3}} I_d$$

Since at $s = -0.5$ the rms rotor current is the same as the rated rotor current, the dc-link current is

$$I_d = \sqrt{\frac{3}{2}} I_{2,\text{rated}} = 329.8 \text{ A}$$

From Eqn (5.30), the slip power is

$$P_{sl} = \left[\frac{3\sqrt{6}}{\pi} \frac{|s|V_1}{m_1} - \frac{3|s|X_{lr}}{\pi} I_d \right] I_d$$

Substituting the values, we get

$$P_{sl} = 42.9 \text{ kW}$$

Since $P_{sl} = |s|P_{ag}$, we get $P_{ag} = 85.8$ kW. Therefore, for slip $s_{fl} = 0.035$ at rated shaft torque,

$$P_{ag,rated} = 3I'^2_{2,rated} \frac{R'_r}{s_{fl}}$$

$$= 92.9 \text{ kW}$$

(a) Since $T_e \propto P_{ag}$, the developed torque under the stated condition is

$$\frac{85.8}{92.9} T_{e,rated} = 0.924 T_{e,rated}$$

(b) At the rated wind speed V_R, $s = -0.5$. Therefore, $w_r = 1.5w_s$. Let the ratio of the rated wind speed to the minimum wind speed be K. To extract maximum power from wind up to this minimum wind speed, the tip speed ratio should remain constant. So, at the minimum wind speed the rotor speed is

$$w_{r,min} = \frac{1.5}{K} w_s$$

Also, with constant tip speed ratio, $T_e \propto w_r^2$. So, at the minimum wind speed,

$$T_{e,min} = \frac{0.924}{K^2} T_{e,rated}$$

The slip at $w_{r,min}$,

$$s_{min} = \frac{w_s - w_{r,min}}{w_s} = \frac{K - 1.5}{K}$$

According to the statement of the problem, the rotor is short-circuited up to $w_{r,min}$. Therefore, with the rotor short-circuited, a shaft torque $T_{e,min}$ will be developed at s_{min}. Assuming the torque slip characteristic of the motor, with the short-circuited rotor, to be linear in the region of interest,

$$T_{e,\text{min}} = \frac{|s|_{\text{min}}}{0.035} T_{e,\text{rated}}$$

s_{min} should be negative, i.e., $K < 1.5$. Therefore,

$$\frac{1.5 - K}{0.035K} T_{e,\text{rated}} = \frac{0.924}{K^2} T_{e,\text{rated}}$$

or $K^2 - 1.5K + 0.03234 = 0$. Since $1 < K < 1.5$, the solution of the above equation gives $K = 1.48$. So the minimum wind speed up to which the maximum power can be extracted from wind is

$$V_{\text{min}} = \frac{1}{K} V_R$$
$$= 0.676 V_R$$

Summary

This chapter addresses grid-connected induction generator and self-excited induction generator operation. The first deals with constant-voltage, constant-frequency output from both squirrel cage and wound rotor induction machines, whose stator windings are directly connected to the grid. The near-synchronous-speed squirrel cage induction generator, driven by a wind turbine via a gear box, prevails dominantly (more than 80%) over the other types of generators in the wind power market. Their manufacturing range extends up to 1.5 MW. Both classical stall and active stall are used with these fixed-speed turbines to limit the power generation at high wind speeds. This system is cheap and simple, but it draws the least amount of energy from wind compared to other technologies for the same wind speed values.

For variable-speed operation, the wound rotor induction machine is used. The stator is directly connected to the grid. The rotor also feeds power to the grid via a converter. The system, known as a double-output induction generator, is the favoured choice for variable-speed, high-capacity turbines in the range 1–4.5 MW. Above the rated wind speed, the output is restricted to rated power by pitching the blades. The system offers good power factor, good speed variation ($\approx 50\%$), and low converter rating ($\approx 35\%$ of the total power). The principle of operation of such a system, its steady-state equivalent circuit model, and the active and reactive power flow control based on the vector control method are presented.

The second issue treated in this chapter is the isolated wind turbine generator operation, where variable-voltage, variable-frequency power is generated by a self-excited induction generator (SEIG) of the squirrel cage type. The self-excitation process, the excitation requirements, the circuit model for SEIG, and the steady-state operation are presented and discussed.

Problems

1. The magnetization curve of a 400-V, four-pole, three-phase, 37.5-kW, 67.6-A squirrel cage induction machine at 1500 rpm is described by the following table.

Phase voltage (V)	Magnetization current (A)
100	5.78
150	8.68
200	12.44
210	13.89
220	16.20
230	19.10
240	23.15
250	28.94
260	36.46
270	46.30

The star-equivalent parameters of the machine, referred to the stator, are $R_s = 0.191$ Ω, $L_{ls} = 1.20$ mH, $R'_r = 0.0812$ Ω, and $L'_{lr} = 1.79$ mH. It is to be operated as a self-excited induction generator feeding a balanced resistive load at the rated voltage and frequency with the stator carrying the rated current. Find the speed of the generator, the load resistance, and the values of the capacitors in the delta connection. Ignore the effect of iron loss.

2. A fixed-pitch wind turbine of radius 8 m with $C_{p,\text{opt}} = 0.4$ at $\lambda_{\text{opt}} = 4.71$ is driving a 400-V, 50-Hz, three-phase, star-connected slip-ring induction generator through a gear

box of ratio 36:1 having an average torque transmission efficiency of 0.96. The induction machine has the following star-equivalent stator-referred parameters: $R_s = 0.355\ \Omega$, $X_{ls} = 0.5\ \Omega$, $R'_r = 0.149\ \Omega$, and $X'_{lr} = 0.750\ \Omega$. The stator is connected to a 400-V, 50-Hz utility network and the slip rings are connected to external resistors. The wind speed is 11 m/s and the turbine is operating at its maximum aerodynamic efficiency. Determine the active power flow into the utility and the resistance of the external resistors. Take the air density as 1.225 kg/m^3 and neglect the iron and frictional losses of the induction machine.

3. A 400-V, 125-kA, 75-kW, three-phase, 50-Hz, four-pole slip-ring induction machine is working as a double-output induction generator under the static slip power control scheme, as shown in Fig. 5.5. The generator shaft is receiving 50 kW of mechanical power at 1580 rpm, and is feeding power to a 400-V bus. The stator-referred parameters of the induction machine and the other system particulars are as follows: $R_s = 0.084\ \Omega$, $R'_r = 0.066\ \Omega$, $L_{ls} = 0.8$ mH, $L'_{lr} = 1.1$ mH, and $L_m = 22$ mH. The dc-link resistance R_f is 0.05 Ω (actual), the stator-to-rotor turn ratio m_1 is 1.8, and the primary-to-secondary turn ratio of the transformer (m_2) is 1.6. How is the delivered power shared by the stator and the rotor under the rated current condition of the machine?

4. A four-pole, three-phase, 150-kW, 400-V, 50-Hz, grid-connected slip-ring induction generator delivers its rated output to the system at 2% slip. At values of slip exceeding 2%, the power delivered by the stator to the system is held constant by controlling the external resistances in the rotor circuit. The induction generator has the following star-equivalent parameters referred to the stator: stator resistance $R_s = 0.022\ \Omega$, rotor resistance $R'_r = 0.0196\ \Omega$, stator leakage reactance $L_{ls} = 0.25$ mH, rotor leakage inductance $L'_{lr} = 0.4$ mH, magnetizing reactance $L_m = 10.51$ mH, and the iron loss is 4875 W. For variable output up to 2% slip and subsequent constant power output up to 10% slip, plot the following quantities against the slip.

(a) External rotor resistance

  (b) Rotor current and stator current
  (c) Stator power output
  (d) Efficiency of the generator

5. The C_p-λ characteristic of a fixed-pitch wind turbine is given by

$$C_p = C_{p,\text{opt}} \frac{\lambda}{\lambda_{\text{opt}}} \left(2 - \frac{\lambda}{\lambda_{\text{opt}}} \right)$$

   in the normal working region. The wind turbine drives a grid-connected squirrel cage induction machine which has a full-load slip of 4% and a breakdown slip of 10%. If the breakdown torque of the machine is 1.5 times the full-load torque, find (i) the furling speed of the turbine as a percentage of the rated wind speed and (ii) the electrical power output of the machine at the furling wind speed in terms of the rated output power. Assume that the gear ratio between the turbine and the machine is chosen such that the machine operates at the rated slip and rated wind velocity with $\lambda = \lambda_{\text{opt}}$. Neglect all mechanical losses and the electrical losses in the machine stator.

6. The grid-connected squirrel cage induction machine of Problem 5 is now replaced by a similarly rated wound rotor induction machine and a generation scheme similar to that shown in Fig. 5.5. The induction machine has a rated full-load slip of 4%. The rotor is short-circuited up to the rated slip and produces constant rated torque beyond the rated slip. If the furling wind velocity is the same as that in Problem 5, find

   (a) the speed of the machine at the furling wind velocity,
   (b) the power output of the generation scheme at the furling wind velocity, and
   (c) the net power factor of the generation system when the rated power factor of the machine is 0.85.

   Assume that the stator-to-rotor turn ratio of the machine and the turn ratio of the line-side transformer are unity. The rotor-side converter is operated at $\alpha = 0$.

7. What should be the turn ratio of a transformer to be connected between the grid and the line-side converter of

the generation scheme of Problem 6 if the line-side converter has a maximum firing angle of 170°? Compare the kVA rating of this line-side converter with that of Problem 6 (no transformer). What will be the net power factor of the system at the furling wind velocity when this transformer is used? Assume this transformer to be ideal.

8. Consider the system shown in Fig. 5.5 for controlling a doubly fed induction generator driven by a wind turbine. The gear ratio between the turbine and the generator is chosen such that the maximum power of the turbine at a specified wind velocity occurs at a slip s_i in the supersynchronous region. With the drop in the wind speed, the converters are controlled to enable the generator to operate over a wide speed range—subsynchronous to supersynchronous—always following the cube-law power characteristic, with the power coefficient remaining at its maximum value. Under such operating conditions, show that in the subsynchronous region the rotor power reaches its maximum at $s = 0.33$. Choose s_i such that the rotor power at this slip equals the maximum rotor power in the subsynchronous region. What percentage of the maximum turbine power is this rotor power? Neglect all losses.

9. The ac output from a self-excited induction generator is rectified by a six-pulse full converter to supply 20 kW to a dc link at 500 V. The induction generator has the same rating and parameters as those given in Problem 1. The generator is driven at 1500 rpm and the voltage at the input to the rectifier is 400 V (line-to-line). Find the value of the capacitance required to sustain this operating condition.

6

Generation Schemes with Variable-speed Turbines

For a given temperature and pressure, the power contained in the wind at a particular site is proportional to the cube of the wind speed. Ideally, the maximum power that a turbine can extract is 0.593, the *Betz coefficient*, times the power contained in the wind. However, the maximum extractable power from a practical turbine is limited to 35–40% of the wind power. For a given turbine, this limit is achievable for a specific ratio of the turbine's rotational speed to the wind speed, which has been discussed in Chapter 1. At other ratios, the turbine output reduces. So, with constant change in wind speed, a natural occurrence, it is desirable for the turbine speed to be adjustable to the wind speed in order to maximize the output—variable-speed generation. Besides increased output, variable-speed turbine operation has many other advantages, in contrast to the constant-speed operation of generators in conventional power generating stations.

6.1 Classification of Schemes

Broadly, four different systems are used for generation of electricity from wind power.

Constant-speed, constant-frequency

The generation scheme in this category is based on fixed-speed technology. The horizontal-axis wind turbine, whose speed can be controlled by using a pitch-control mechanism, operates at a constant speed and drives, through a gear box, a synchronous or an induction generator that is connected to the power network.

A constant-speed wind turbine can achieve maximum efficiency at the speed that gives the tip speed ratio the value corresponding to the maximum power coefficient $C_{p,\text{opt}}$. Its main weakness lies in its poor energy capture from the available wind power at other wind speeds. Moreover, a pitch-control mechanism adds considerably to the cost of the machines and stresses the operating mechanism and the machines.

Near-constant-speed, constant-frequency

In this scheme, induction generators feed power to the utility network at variable slip. Here also the generators are driven by horizontal-axis wind turbines but with a less stringent pitch angle controller, which can maintain small values of slip.

Variable-speed, variable-frequency

This scheme employs capacitor self-excited three-phase or single-phase induction generators for small-scale power generation as a source of isolated supply to feed frequency-insensitive loads.

Variable-speed, constant-frequency

Wind turbines are basically variable-speed prime movers. This category implies a wide and continuous range of variable-speed operation of the turbine and the processing of power ultimately at the synchronous frequency of the utility system. Variable-speed operation of wind turbines offers several benefits. So there is a general trend now towards generation schemes employing variable-speed turbines. There are many reasons for such a choice, which may be briefly summarized as follows.

(a) Continuous operation of wind turbines at the optimum tip speed to wind speed ratio by changing the rotor speed with the wind velocity. This increases energy capture even under low wind conditions.

(b) Reduction in noise emission from wind turbines at low wind speeds.

(c) Reduction in the size and weight of the gear box, or its total elimination, together with the associated noise

(d) The possibility of power smoothing due to the inertial energy storage in the turbine rotor as the wind gusts above the average level. With reduction in the wind speed, the power flow level in the network can be maintained by deriving additional energy from the inertia of the system. The time trace of the power output of a constant-speed wind turbine is characterized by high-frequency fluctuations superimposed on slow power variations owing to the short-term wind fluctuations and inherent time lag in the wind turbine control system. On the other hand, the time trace of the power output of a variable-speed system is considerably smoother due to the rotor flywheel effect.

The variable shaft speed leads to variable-voltage, variable-frequency output from the generator, in general. However, in certain systems, the output voltage magnitude can be maintained constant, or within a range, by a voltage-regulating system.

The variable-voltage, variable-frequency system requires efficient power electronic ac/dc/ac converters for interfacing with the utility system. Converters using power electronic devices have good dynamic performance, and can provide high-quality sine wave current in the generator and the power network. They can also help to control the real as well as the reactive power of the system. Furthermore, when a number of wind generators operate in parallel, the converters can optimize the output of each machine in order to increase the total power output, by allowing different machines to operate at different speeds.

6.2 Operating Area

Though maximum energy capture is an important factor in the variable-speed operation of wind turbines, in actual operation, limits on the shaft torque, power, and speed determine the overall control strategy. The operating area of a wind turbine is depicted in Fig. 6.1, showing the torque–speed characteristics with increasing wind velocity. AB, BD, and DE, respectively, define the limits of shaft torque, power, and speed. Within the area $OABDE$ a number of control strategies can be adopted. The operating point in the torque–speed plane for variable-speed operation at maximum C_p follows the square-law curve OC. At C, the power limit is reached, beyond which the turbine operates along CD, maintaining a constant power limit at reduced torque until the speed limit Q_1Q_2 is reached. Constant-speed operation along P_1P_2 and Q_1Q_2 are limited by the maximum torque and maximum power, respectively. If the torque and the speed constitute the only turbine operation limits, $OAZE$ would be the *safe operating area* of the turbine. This will allow an improvement in the energy capture up to W.

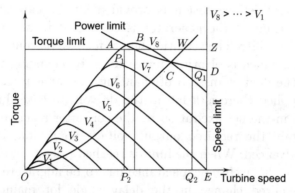

Fig. 6.1 Wind turbine torque–speed characteristics at different wind speeds

6.3 Induction Generators

Self-excited generators working in isolation with variable-speed prime movers, such as wind turbines, have poor voltage and frequency regulation. Generally, squirrel cage induction machines are chosen due to their ruggedness, low cost, and low maintenance. For frequency-insensitive loads, such as heating and lighting, it is adequate to maintain a near-constant terminal voltage. In fact, irrespective of the nature and amount of load, a constant terminal voltage with admissible regulation is required in most applications. The generated ac voltage may either be used directly or converted into dc voltage. Dc power can be used directly in certain dc equipment, such as battery chargers, or fed to the ac mains, or load, through an inverter.

A number of methods are available for regulating the voltage of the self-excited induction generator, each with their advantages and limitations. The application of power semiconductor devices, controlled converter circuits, and control algorithms have resulted in suitable regulating schemes for self-excited variable-speed squirrel cage generators.

6.3.1 Cage Rotor Induction Generator

Controlled firing angle scheme with ac-side capacitor

A method to obtain dc power at a controllable voltage from a variable-speed generation system is shown in Fig. 6.2. Initially, the necessary self-excitation is provided by the capacitor bank and the output of the generator is connected to a three-phase controlled rectifier. The attention here is focused on controlling the dc voltage, which is of prime importance, by changing the firing angle of the devices instead of changing the capacitors. When a load is applied there is a fall in the dc voltage level due to the reduction in the net excitation by the converter reactive current. To maintain the required output voltage regulation, the firing angle is advanced. When the limit is reached (ideally diode-bridge operation), the exciting capacitance has to be augmented.

The required change in the delay angle for maintaining a constant dc voltage is very sensitive to the speed change. At a higher speed, when the no-load voltage rises sharply, a large delay angle is required for maintaining a constant dc voltage. With a fixed excitation capacitor, the useful speed range for a constant dc voltage, employing a line-commutated controlled rectifier, is extremely restricted. The lower end of the speed range is limited by the minimum excitation speed and the upper end by the excessive rise in the induced voltage and the large delay angle required. In order to increase the speed range, the capacitance value needs to be changed. In this regard the scheme is not self-contained. To obtain a wide speed range, two or more sets of capacitor banks are needed. Therefore, all the basic problems of a self-excited induction generator are partially remedied, but remain unsolved.

For the same excitation capacitance, the useful speed range can be increased if a forced-commutated rectifier is used in place

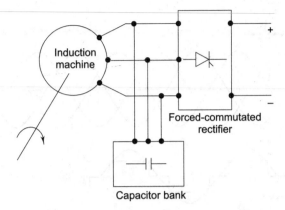

Fig. 6.2 Self-excited induction generator with an ac-side capacitor and a forced-commutated rectifier

of a line-commutated rectifier. The basic principle is enunciated by the voltage and the current waveforms shown in Fig. 6.3. It assumes sinusoidal converter input voltages, ripple-free output current, and ideal switching elements. Advancing the firing angle (i.e., negative delay angle) produces a capacitive effect at the machine terminals since the rectifier input current now leads the corresponding voltage. This is in tune with the requirements of a self-excited induction generator, where the effective excitation capacitance must increase with increasing load current. At high speeds and light loads, the firing angle can be delayed (i.e., α made positive) so that the rectifier appears as an equivalent RL load to reduce the effective VAR supplied to the machine, and, thereby, avoid high voltage at the machine terminals.

For ripple-free rectifier output current I_d, the rms value of the fundamental component of the rectifier input current is

$$I_1 = \frac{1}{\sqrt{2}} \frac{2\sqrt{3}}{\pi} I_d \tag{6.1}$$

If the phase-a voltage is given by

$$v_a = \frac{\sqrt{2}V}{\sqrt{3}} \sin \omega t \tag{6.2}$$

then the instantaneous value of the fundamental component of the phase-a current can be expressed as

$$i_{a1} = \sqrt{2} I_1 \sin(\omega t - \alpha) \tag{6.3}$$

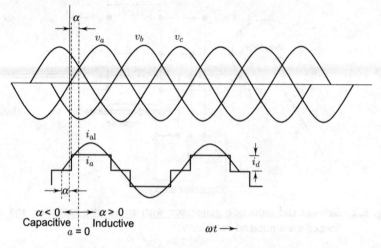

Fig. 6.3   Input voltage and current waveforms

Equating the active power on both the sides of the rectifier, we get

$$\sqrt{3}V I_1 \cos \alpha = V_d I_d \tag{6.4}$$

Substituting the expression for I_1 in Eqn (6.4) yields the following relation between the average dc output voltage and the line-to-line ac input voltage.

$$V = \frac{\pi V_d}{3\sqrt{2}\cos \alpha}$$

The per-phase rms voltage in terms of the dc output voltage is

$$V_p = \frac{\pi V_d}{3\sqrt{6}\cos \alpha} \tag{6.5}$$

From Eqns (6.1) and (6.3), the rectifier input current can be written as

$$i_{a1} = \sqrt{2}I_1 \cos \alpha t \sin \omega t - \sqrt{2}I_1 \sin \alpha \cos \omega t$$

$$= \frac{2\sqrt{3}}{\pi} I_d \cos \alpha \sin \omega t - \frac{2\sqrt{3}}{\pi} I_d \sin \alpha \cos \omega t$$

The first term represents the in-phase active component of the phase-a current in phase with v_a and the second term represents the reactive current, phase-shifted by $-90°$ from v_a.

Therefore, the rms active and reactive components of the rectifier input currents are, respectively,

$$I_p = \frac{\sqrt{6}}{\pi} I_d \cos \alpha$$

and

$$I_q = \frac{\sqrt{6}}{\pi} I_d \sin \alpha \tag{6.6}$$

The star-connected equivalent, i.e., the reflected, per-phase shunt resistance and reactance across the machine terminals are obtained by dividing the phase rms voltage in Eqn (6.5) by the rms values of the current in Eqn (6.6). They are given by

$$R_{eq} = \frac{\pi^2 R}{18 \cos \alpha^2}$$

and

$$X_{eq} = \frac{\pi^2 R}{9 \sin 2\alpha} \tag{6.7}$$

For $\alpha > 0$, X_{eq} is the inductive reactance, and for $\alpha < 0$, X_{eq} becomes the capacitive reactance. So, when the load current is increased by decreasing the value of R, X_{eq} decreases. At the same time, the capacitive current for $\alpha < 0$ increases, preventing the terminal voltage from dropping. When the load current is too low and/or the speed is high, the system may be required to operate with positive α. A simplified circuit model of the generator/rectifier, referred to the stator and normalized to the base frequency, is shown in Fig. 6.4. The per-unit value of the frequency F can be obtained from the power balance equation as follows:

$$\frac{E_p^2(F - v)}{R_r} + V_p I_p + R_s(I_p^2 + I_m^2) = 0 \tag{6.8}$$

The behaviour of the system in terms of the relationship between the firing angle and the speed required to provide a constant dc voltage and supply a fixed load resistance can be deduced by an iterative process using the above equations in conjunction with the circuit equations and the magnetization characteristic.

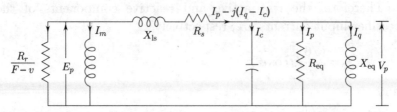

Fig. 6.4 Circuit model of a self-excited induction generator with a controlled rectifier

Inverter/converter system with dc-side capacitor

In conventional self-excited variable-speed induction generators with a bank of capacitors connected across the machine terminals, the value of capacitance required is almost inversely proportional to the square of the prime mover speed, calling for impracticably large values of capacitance at low speeds. Even then, further stages of power conversion are needed to feed conventional dc/ac loads due to the variable magnitude and frequency of the generated voltage. These problems can be overcome to a large extent by using a power converter with a large dc-side capacitor connected between the induction generator and the power network. This power conversion equipment comprises two fully controllable pulse width modulated voltage source inverters to provide independent control of the net generator current. Both of them have bidirectional sinusoidal operation. The power flow into and out of bridge 1 and the voltage of the dc link can be controlled by controlling the phase and the magnitude of the modulating signal relative to the network voltage. This also enables one to derive power from the network to start the turbine in certain wind generator configurations, such as the Darrieus turbine. Bridge 2 can be controlled by the conventional scalar variable-voltage/variable-frequency method or by the flux vector control scheme. These schemes are outlined below with reference to a dc load.

The scalar method Figure 6.5 shows the control scheme of an induction generator–static inverter system with a dc load and capacitor. The frequency of the inverter is adjusted to give a small negative slip. Assuming that the capacitor voltage builds up to a steady value, the working of the system can be described as follows. The PWM inverter converts the dc voltage into ac voltage, which supplies the necessary magnetizing ampere-turns to establish the air-gap flux of the machine. If the slip is made negative, the mechanical power gets converted into electrical power, which, through the feedback diodes of the inverter, supplies the load connected across the capacitor. At the same time, it also replenishes the charge that has been drained from the capacitor.

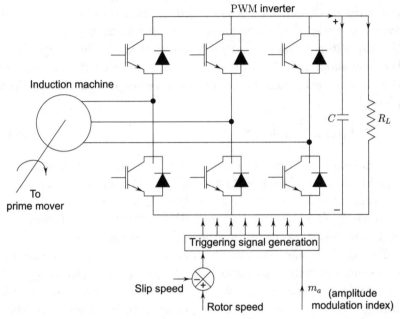

Fig. 6.5 Self-excited induction generator with a dc-side capacitor and a converter/inverter system

An increase in the load tends to decrease the capacitor voltage, which can be compensated by increasing the slip, i.e., reducing the inverter frequency. Again, in order to maintain a nearly constant dc voltage at reduced prime mover speed, it is necessary to reduce the modulation index (m_a) of the inverter in proportion to the speed. Thus, to have a fairly constant dc voltage for varying loads and prime mover speeds, the slip as well as m_a should be controlled. Although the output of the system is dc, its near constant value makes it easier both for supplying it to dc loads and for feeding it to an ac grid system through an inverter.

The system has several advantages: (a) a single large dc capacitor is sufficient, instead of several banks of ac capacitors, (b) the value of the dc capacitor is not critical provided it is sufficiently large, and (c) the output voltage can be maintained constant over a wide range of speeds.

The analysis of the system is based on the assumption that the build-up time of the capacitor voltage from a low initial value is large compared to the period of the alternating cycle. As a result, the inverter output can be assumed to have constant amplitude over one cycle of the generated voltage waveform. Let $V_c(t)$ be the capacitor voltage at any instant during the build-up. Since the sinusoidal PWM modulation technique is used, the peak of the fundamental component of the inverter output voltage per phase is

$$V_1(t) = \frac{1}{2} m_a V_c(t) \tag{6.9}$$

For maintaining constant air-gap flux, the machine terminal V/f should be almost constant. Therefore, m_a is made proportional to the generated frequency for a constant dc voltage.

Since $V_c(t)$ is assumed to remain constant at least over one cycle, the inverter output for different phases can be assumed to be balanced. This suggests that we can model the induction machine in a synchronously rotating frame with the d-axis coinciding with the rotating space voltage vector of magnitude V_1. The d-q axis voltage equations in the synchronously rotating frame can then be written as

$$
\begin{bmatrix} V_1 \\ 0 \\ 0 \\ 0 \end{bmatrix}
$$

$$
= \begin{bmatrix}
R_s + L_s p & -\omega_e L_s & Mp & -\omega_e M \\
\omega_e L_s & R_s + L_s p & \omega_e M & L_s p \\
Mp & -(\omega_e - \omega_r)M & R_r + L_r p & -(\omega_e - \omega_r)L_r \\
(\omega_e - \omega_r)M & Mp & (\omega_e - \omega_r)L_r & R + L_r p
\end{bmatrix}
$$

$$
\times \begin{bmatrix} I_{ds} \\ I_{qs} \\ I_{dr} \\ I_{qr} \end{bmatrix} \tag{6.10}
$$

where I_{ds}, I_{qs} and I_{dr}, I_{qr} are the direct-axis, quadrature-axis components of the time-varying peaks of the stator phase and rotor phase currents, respectively.

As the saturation of the main magnetic circuit is a critical factor, the magnetizing inductance should be estimated at each stage of the computation from the magnetizing current, whose peak value is given by

$$I_m = \sqrt{(I_{ds} + I_{dr})^2 + (I_{qs} + I_{qr})^2} \tag{6.11}$$

The value of M is generally obtained from experimental data in the form

$$M = f(I_m)$$

Because of the negative slip, the input current I_1 to the machine has a negative real part while the quadrature component lags behind the voltage. Viewed as a generator, the current output of the machine is $(-I_1)$, which leads the voltage by an angle, say, ϕ. The feedback diodes in the inverter help rectify the three-phase output currents. The rectified current, which is pulsating in nature with a frequency of $6f_s$, feeds the dc load connected across the capacitor. Depending on the value of C, the pulsating current causes a corresponding ripple in the voltage riding above the mean value. The average voltage as well as the ripple go on increasing as long as the electrical power developed by the generator exceeds the sum total of the power consumed by the load and the machine winding copper losses.

The rectified current (dc-link current), during any one cycle, has the waveform shown in Fig. 6.6 for an inverter without modulation. With modulation, there are zero intervals in the link current, reducing its value. The Fourier series expansion of the current, neglecting harmonics higher than $6\omega_s$, is

$$I_{dc} = -m_a \left[\frac{3}{\pi}(I_1 \cos \phi) + \frac{36}{35\pi} I_1 \sin \phi \sin(6\omega_s t) \right.$$
$$\left. - \frac{36}{35\pi} I_1 \cos \phi \cos(6\omega_s t) \right]$$

As the d-axis has been taken to coincide with the voltage space-vector, the in-phase $(I_1 \cos \phi)$ and quadrature $(I_1 \sin \phi)$

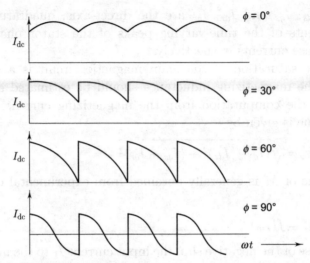

Fig. 6.6 Dc-link current waveform for an unmodulated inverter for different power-factor angles

components of the machine current are I_{ds} and I_{qs}, respectively. Since the current is used in feeding the load and charging the capacitor, the above equation can be written as

$$C\frac{dV_c}{dt} + \frac{V_c}{R_L} = -m_a\left[\frac{3}{\pi}I_{ds} + \frac{36}{35\pi}I_{qs}\sin(6\omega_s t)\right.$$
$$\left. - \frac{36}{35\pi}I_{ds}\cos(6\omega_s t)\right] \qquad (6.12)$$

Starting with a low initial voltage on the capacitor and zero machine currents, Eqns (6.9)–(6.12) can be solved step by step for a given rotor speed, slip, modulation index, dc capacitance and load resistance until the machine voltage builds up to a steady value. At each step, M is to be updated suitably, depending on the magnetizing current.

The flux-vector scheme Field-oriented techniques have made possible the development of high-performance ac drives over a large speed range, and offer independent control on the torque and the flux. As in the case of drives, the field-oriented principle can equally be applied to improve the dynamic performance of a variable-speed wind turbine generator set with independent control on both the active and the reactive power through PWM current-regulated converters.

The rotor-flux-oriented system, a very familiar approach in drives, is obtained by accepting the machine rotor flux vector as the *d*-axis. It provides excellent decoupled control when the correct rotor parameters are used in the control algorithm. The parameter sensitivity of the rotor-flux-oriented system affects both the dynamic and the steady-state performances of the system when the rotor parameters are not correctly estimated. These effects become more severe in large machines and in machines operated at reduced flux levels. On the other hand, the low parameter sensitivity of the stator-flux-orientation technique represents an important feature. Incorrect values of the rotor parameters affect the transient response but have hardly any significant effect on the steady-state performance. Orientation to the stator flux offers the advantages of robust flux estimation and direct control of the stator voltage/current, an important aspect in a generation system.

In field orientation with stator flux as the feedback quantity, the machine model must first be expressed in a reference frame oriented to the stator flux.

For the orientation of the *d*-axis to the stator flux,

$$\lambda_{ds}^e = \lambda_s, \quad \lambda_{qs}^e = 0 \tag{6.13}$$

where the superscript *e* implies a synchronously rotating *d-q* axis in the stator-flux-oriented system. Eliminating the rotor currents from Eqn (3.54) using Eqn (3.53) and using the condition in Eqn (6.13), we have

$$\lambda_{dr}^e = \frac{L_r}{M}(\lambda_s - \sigma L_s i_{ds}^e)$$

$$\lambda_{qr}^e = -\frac{L_r}{M}\sigma L_s i_{qs}^e \tag{6.14}$$

where $\sigma = 1 - M^2/L_r L_s$ is the total leakage coefficient. Using Eqns (6.13), (6.14), and (3.53) in Eqns (3.51) and (3.52) yields

$$(1 + \tau_r p)\lambda_s = -\omega_{sl}\tau_r \sigma L_s i_{qs}^e + (1 + \sigma\tau_r p)L_s i_{ds}^e \tag{6.15}$$

and

$$(1 + \sigma\tau_r p)L_s i_{qs}^e - \omega_{sl}\tau_r(\lambda_s - \sigma L_s i_{ds}^e) = 0 \tag{6.16}$$

where $\omega_{sl} = \omega_e - \omega_r$ is the slip speed in electrical rad/s, obtained from Eqn (6.16) as follows:

$$\omega_{sl} = \frac{(1 + \sigma\tau_r p)L_s i_{qs}^e}{\tau_r(\lambda_s - \sigma L_s i_{ds}^e)} \tag{6.17}$$

The elimination of ω_{sl} from Eqns (6.15) and (6.16) gives

$$\frac{(1 + p\tau_r)\lambda_s}{L_s(1 + \sigma\tau_r p)} = i_{ds}^e - \frac{\sigma L_s i_{qs}^{e2}}{\lambda_s - \sigma L_s i_{ds}^e} \tag{6.18}$$

This is what happens inside the machine. As i_{ds}^e is oriented in the direction of the flux vector, it is desirable for i_{ds}^e to decide λ_s. Equation (6.18) shows a coupling between λ_s and i_{qs}^e. In order to nullify the effect of i_{qs}^e, the output from the flux regulator in the control scheme is augmented by i_{dq}^{e*}, as depicted in Fig. 6.8, to have the command value

$$i_{ds}^{e*} = G_d(p)(\lambda_s^* - \hat{\lambda}_s) + i_{dq}^{e*} \tag{6.19}$$

where

$$i_{dq}^{e*} = \frac{\sigma L_s (\hat{i}_{qs}^e)^2}{\hat{\lambda}_s - \sigma L_s \hat{i}_{ds}^e} \tag{6.20}$$

and $G_d(p) = K_P + K_I/p$ is the transfer function of the proportional-integral-type flux controller. Equation (6.20) describes a decoupler. Substituting Eqn (6.19) in Eqn (6.18) gives

$$\frac{(1 + p\tau_r)\hat{\lambda}_s}{L_s(1 + \sigma\tau_r p)} = G_d(p)(\lambda_s^* - \hat{\lambda}_s) \tag{6.21}$$

which is independent of i_{qs}^e. The flux controller decides the machine flux.

From Eqn (6.18) it is evident that for a machine with low leakage factor, stator flux vector control permits a very close decoupling of the excitation control of the induction machine with its active power control.

Using the condition stated in Eqn (6.13) in Eqns (3.49) and (3.50) for a constant value of λ_s gives

$$v_{ds}^e = R_s i_{ds}^e$$

$$v_{qs}^e = R_s i_{qs}^e + \omega_e \lambda_s$$

Since the per-unit value of R_s is very small, the machine terminal voltage is practically decided by v_{qs}^e. The stator space voltage vector and the stator flux vector are very approximately in quadrature, with the voltage vector leading $\vec{\lambda}_s$. The stator power from Eqn (3.64) can practically be equated to

$$P_s = v_{qs}^e i_{qs}^e$$

In a power generation scheme with ac/dc systems, the dc link is associated with active power. The q-axis current regulates the active power supplied to the dc link. Compared with the field-oriented induction motor drive system, in the generator operation, speed control is replaced by dc-link voltage control to obtain the command value for the q-axis component of the stator current vector. The d-axis current is associated with the flux, and represents the reactive power demand of the generator.

To obtain the reactive component of the reference current and proper orientation of the current vector, it is necessary to develop a procedure to determine the amplitude and the space phase angle of the stator flux vector from easily measured quantities such as voltage, current, and speed.

Writing Eqns (3.49) and (3.50) in terms of the stationary reference frame, i.e., for $\omega_e = 0$,

$$\lambda_{ds}^s = \int \left(v_{ds}^s - R_s i_{ds}^s \right) dt$$

and

$$\lambda_{qs}^s = \int \left(v_{qs}^s - R_s i_{qs}^s \right) dt$$

The stator flux vector is given by

$$\vec{\lambda}_s = \lambda_{ds}^s + j\lambda_{qs}^s = \lambda_s e^{j\theta} \tag{6.22}$$

Figure 6.7 presents the spatial relations between the stator-flux-oriented reference frame and the stationary reference frame attached to the a-phase winding. Once θ is estimated from Eqn (6.22), the command values of the phase-winding currents are determined from Eqn (3.47).

Figure 6.8 presents the basic power circuit and the block schematic diagram of the control scheme for the generator. The

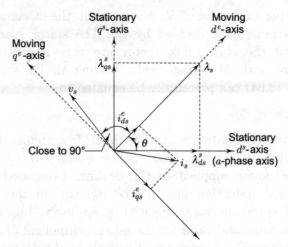

Fig. 6.7 Components of the stator flux and current vectors in the stationary frame and the stator flux frame

voltages and currents of the induction generator are measured and
transformed into two-axis values in the stationary reference frame
to feed the flux vector calculator. The amplitude and the phase
angle θ of the flux vector are calculated using Eqn (6.22). The
flux controller sets the direct-axis component i_{ds}^{e*} of the current
vector and operates on the difference between the calculated stator
flux and the stator flux reference. Similarly, the dc voltage error
serves as an input to the voltage controller to determine the
reference value of the quadrature component of the current vector
for adequate power flow to the dc link. The dc reference voltage
is constant up to a base speed, after which it decreases, and

$$V_{\text{dc}} > \frac{\sqrt{6}}{2}\pi p f_r \lambda_s$$

The two-axis rotating current reference values are converted into
stationary two-axis reference current values by the space angle θ of
the flux vector. A stationary two- to three-phase transformation
then yields three-phase reference current values. The hysteresis
controller compares the three-phase reference currents with the
actual stator currents and operates the inverter in such a way as
to confine the generated current within a certain band around the
reference values.

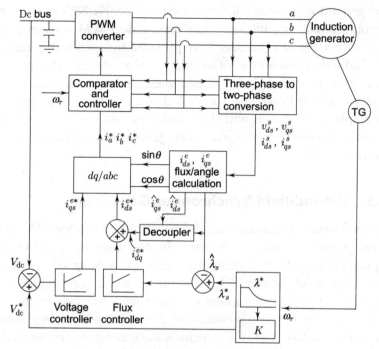

Fig. 6.8   Configuration of the stator-flux-oriented control system

6.4   Doubly Fed Induction Generator

The wound rotor induction machine, commonly known as the doubly fed induction generator, is finding increasing application, particularly in the megawatt range, in variable-speed wind energy conversion systems. When compared with motoring operation, the power handling capability of a wound rotor induction machine as a generator theoretically becomes nearly double. The rotor of the generator is coupled to the turbine shaft through a gear box so that a standard (1500/1800 rpm) wound rotor induction machine can be used. The gear ratio is so chosen that the machine's synchronous speed falls nearly in the middle of the allowable speed range of the turbine (nearly 60–110 %). Above the rated wind speed, power is limited to the rated value by pitching the blades. The stator is directly connected to the fixed-frequency utility grid while the rotor collector rings are connected via back-to-back PWM voltage source inverters and a transformer/filter to the same utility grid. As the rotor power is a fraction of the total power of the generator, a rotor converter rating of nearly 35% of the rated turbine power is sufficient. The rotor-side PWM converter is a stator flux based controller that provides independent control of the induction machine's active and reactive powers. The grid-side converter is the dc-link voltage regulator that enables power flow to the grid, keeping the dc-link voltage level constant. The principle of operation with the double-output system coupled to a variable-speed turbine has been presented in Chapter 5 (see Section 5.1.5).

6.5 Wound-field Synchronous Generator

A synchronous generator can also be employed in a variable-speed wind energy conversion scheme. In the case of a self-excited squirrel cage induction generator under variable-speed operation, the amount of lagging VAR required for voltage regulation is supplied by switched capacitors, controlled rectifiers with fixed capacitors, or frequency-controlled generator-side voltage source inverters. In the case of a synchronous generator, the generated voltage can be controlled more easily by using a voltage regulator in the field circuit. It requires no shunt capacitors/controlled inductors to achieve voltage regulation.

Figure 6.9 illustrates a power generation scheme at the grid frequency, employing a variable-speed synchronous generator with excitation control. The conversion scheme for the variable output of the synchronous generator consists of an uncontrolled diode-bridge rectifier, a filter capacitor, a smoothing dc-link choke, and a line-commutated thyristor bridge inverter. The diode and the thyristor bridges generate harmonic currents and voltages, and their interaction results in considerable power fluctuation through the dc link. This interaction is considerably reduced by means of the capacitor filter at the output of the diode bridge. The control strategy is based on the speed cube law for optimal power output from the wind turbine.

The excitation emf of the generator is proportional to the product of the field current I_f (assuming magnetic linearity) and the speed ω. If the dc-link current I_{dc} and the field current I_f are

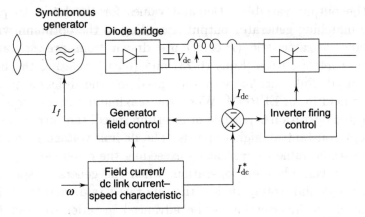

Fig. 6.9 Control scheme for a synchronous generator with variable-speed drive and utility interface

made proportional to speed, i.e.,

$$I_{dc} = k_{dc}\omega \qquad (6.23)$$

and

$$I_f = k_f\omega \qquad (6.24)$$

the excitation emf will be proportional to ω^2 and the fundamental component of the generator current will be proportional to ω. The generated power will be less than that given by the cubic power versus speed characteristic $(P \propto \omega^3)$ owing to the synchronous reactance of the machine, which causes a phase difference between the excitation emf and the fundamental component of the stator current. The output power of the synchronous generator in the dc link is given by

$$P_{dc} = V_{dc}I_{dc}$$

which is the generated power minus the stator copper loss and the diode-bridge loss. Therefore, in order to make P_{dc} follow the cubic law in speed, the field current is required to have a component dependent on V_{dc}.

In the system illustrated in Fig. 6.9, the turbine (or the generator) speed and the dc-link voltage are chosen as the input variables, and the dc-link current and the generator field current as the output variables. Demand values for the latter to provide matching generator output according to the optimum wind turbine characteristic $(P \propto \omega^3)$ are drawn from pre-estimated linear control curves [Eqns (6.23) and (6.24)], relating the field current/dc-link current with the speed for the different power levels held in an EPROM. When the maximum current required for rated operation is reached at a certain speed, the current value is kept constant at higher speeds. The dc-link voltage is limited to a specific value to prevent its exceeding the rated value.

Under variable-speed operation, when the generator speed is low, the dc-link voltage is less than its rated value and the firing angle of the inverter has to be advanced in order to meet the generated power, resulting in a reduced power factor at the grid. The power factor can be improved by using a boost chopper and a PWM inverter as shown in Fig. 6.10. Since the machine inductances can be utilized and the filter capacitor is already present, the additional components required for the boost chopper are a transistor and a forward diode connected before the filter capacitor.

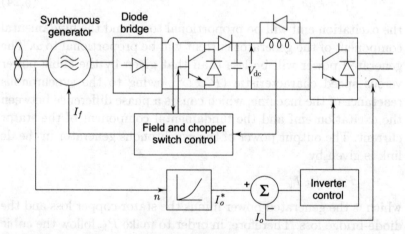

Fig. 6.10 Variable-speed synchronous generator control with a boost rectifier

A line-commutated, six-pulse phase-controlled inverter returns the generated energy to the network with a distorted current waveform, rich in low-order harmonics (5th and 7th), at varying power factor. To deal with this problem, one can use a PWM voltage source inverter, which provides the interface with the network to return power with near sine wave quality current at any desirable power factor. For unity power factor operation, the output current of the inverter is proportional to the power, and hence the cube law with speed can be the guiding factor to decide on the set value of the inverter output current. With increase in the power demand, the dc-link voltage tends to fall while the generator-side step-up converter tries to maintain a constant dc-link voltage by controlling the power flow from the generator. The control of the step-up chopper is based on the hysteresis switch sensing the output at the diode-bridge rectifier and the speed of the generator.

The use of the PWM rectifier, illustrated in Fig. 4.26, in place of the diode-bridge rectifier or the boost rectifier, has several benefits. It reduces the level of current harmonics in the generator, with the associated reduction of iron and copper losses in the generator, and facilitates the adjustment of the power factor.

6.6 The Permanent Magnet Generator

The basic wind energy conversion requirements of a variable-speed permanent magnet generator are almost the same as those for a wound-field synchronous generator. The power circuit topology is basically the same as that shown in Fig. 6.9. The permanent magnet generator dispenses with the need for external excitation. Therefore, the output voltage, under variable-speed operation, varies both in frequency and in magnitude. As a result, the dc-link voltage changes in an uncontrolled manner. The control, therefore, is realized via the dc/ac converter, i.e., the inverter, on the grid side.

The turbine shaft speed is controlled by the opposing reaction torque of the generator. The optimum values of the dc voltage and current can be obtained as follows. With the diode bridge, assuming ripple-free dc-link current, the generator terminal voltages and currents are related to those of the dc-link circuit is follows:

$$V_g = \frac{\pi}{3\sqrt{6}} V_d \qquad (6.25)$$

and

$$I_g = \frac{\sqrt{6}}{\pi} I_d \qquad (6.26)$$

We can see from Eqn (3.109) and (6.25) that for any particular dc-link voltage, there is a minimum speed below which there is no generation. For any wind speed, there exists an optimum turbine speed at which maximum power can be extracted from the wind. This optimum power versus speed characteristic follows the law

$$P_{\text{opt}} = K_{\text{opt}} \omega_r^3 \qquad (6.27)$$

and

$$T_{p,\text{max}} = K_{\text{opt}} \omega_r^2 \qquad (6.28)$$

Equating Eqn (6.28) to Eqn (3.111) at shaft speed ω_r, for maximum turbine power we have

$$K_{\text{opt}} \omega_r^2 = T_{\text{max}} \sin 2\delta \qquad (6.29)$$

Since δ is known from Eqn (6.29) for a given shaft speed ω_r, the optimum dc-link voltage and current for maximum energy capture can be determined from the successive use of Eqns (3.106), (3.109), (3.107), (6.25), and (6.26). The inverter is controlled in such a manner that this optimum current drawn from the dc link occurs at this optimum voltage.

The power flow from the voltage source inverter to the grid is given by

$$P_I = 3 \frac{V_S V_I}{X_L} \sin \delta \qquad (6.30)$$

where δ, the power angle, is the phase lead of the inverter output voltage V_I with respect to the grid voltage V_S, and is controlled directly by the timing of the commencement of the device gating signal referenced to the grid voltage. The inverter output voltage V_I is a function of the dc-link voltage and the device switching pattern, i.e.,

$$V_I = K_I V_d \qquad (6.31)$$

With negligible inverter loss and under steady-state operation

$$P_I = V_d I_d \qquad (6.32)$$

From Eqn (6.30), (6.31), and (6.32),

$$I_d = 3\frac{V_S K_I}{X_L} \sin \delta \qquad (6.33)$$

Specific values of V_d and I_d are decided by the desired energy capture for a given turbine shaft speed. According to Eqn (3.114), for given values of V_S and X_L, the control of reactive and active power flow are effected by K_I and δ. If P^* and $\cos \phi^*$ are the desired real power and the power factor, respectively, the desired reactive power Q^* is given by

$$Q^* = P^* \tan \phi^* \qquad (6.34)$$

From Eqns (3.113) and (3.114),

$$\delta^* = \arctan\left(\frac{P^*}{Q^* + V_S^2/X_L}\right) \qquad (6.35)$$

and

$$V_I^* = \frac{P^* X_L}{3V_S \sin \delta^*} \qquad (6.36)$$

From Eqn (6.31)

$$K_I^* = \frac{V_I^*}{V_d^*} \qquad (6.37)$$

Equations (6.36) and (6.37) decide the control variables for the inverter control scheme for the generator system shown in Fig. 6.11. Inputs to the controller are the turbine shaft speed that sets the reference power, grid voltage, desired power factor, and the dc-link voltage corresponding to the speed. The outputs are

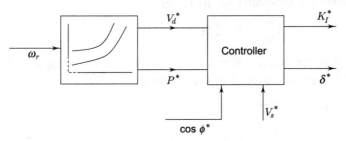

Fig. 6.11 Generation of the reference power angle and conversion factor

the power angle (δ) and the dc-to-ac conversion factor (K_I) for controlling the inverter.

Worked Examples

1. A 440-V, 250-A, 120-pole, 50-Hz, three-phase permanent magnet motor has a reactance of 0.3 Ω per phase. The generated emf per phase is given by

$$E_g = 1.05\omega$$

where ω_s is the angular frequency of the rotor. The generator is driven by a wind turbine whose optimum power as a function of the speed of rotation (in rpm) is given by

$$P_{\text{opt}} = 1.38N^3$$

The generator feeds power to a dc link through a diode-bridge rectifier. Find the optimal dc-link current–voltage characteristic for a rotational speed range of the 30% to 100% of the rated speed. Ignore transmission losses.

Solution

The machine inductance is

$$L_s = \frac{0.3}{2\pi50} = 0.955 \text{ mH}$$

Let the speed be 40 rpm. The frequency generated at 40 rpm is

$$f_{40} = \frac{40 \times 120}{120} = 40 \text{ Hz}$$

The machine reactance at 40 Hz is

$$X_s = 0.3\frac{40}{50} = 0.24 \text{ }\Omega$$

Given the optimum power versus speed (rpm) relationship $P_{\text{opt}} = 1.38N^3$, the corresponding torque relation is

$$T_{p,\text{max}} = \frac{P_{\text{opt}}}{\omega_r}$$

$$= \left(1.38\frac{60}{2\pi}\right)N^2$$

$$= 13.18N^2$$

$$= 1203\omega_r$$

The maximum generating torque that can be developed [see Eqn (3.112)],

$$T_{\max} = \frac{3}{4} \frac{120 \times 1.05^2}{0.955 \times 10^{-3}}$$

$$= 103,900 \text{ N m}$$

From Eqn (6.29),

$$13.18 \times 40^2 = 103{,}900 \sin 2\delta$$

Therefore,

$$\sin 2\delta = 0.203$$

i.e.,

$$\delta = 5.85°$$

From Eqn (3.106), the excitation emf is obtained as

$$E_g = 1.05 \times 2\pi \times 40$$

$$= 263.9 \text{ V}$$

Consequently, the generator terminal voltage [Eqn (3.109)] is

$$V_g = 263.9 \cos 5.85° = 262.5 \text{ V}$$

From the phasor diagram in Fig. 3.23,

$$\frac{I_g X_s}{E_g} = \sin \delta$$

i.e.,

$$I_g = \frac{263.9 \times \sin 5.85}{0.24}$$

$$= 112 \text{ A}$$

The dc-link voltage and current [as per Eqns (6.25) and (6.26)] are

$$V_d = \frac{262.5 \times 3 \times \sqrt{6}}{\pi} = 614 \text{ V}$$

$$I_d = \frac{112 \times \pi}{\sqrt{6}} = 143.6 \text{ A}$$

So the dc-link power is

$$V_d I_d = 614 \times 143.6 = 88.2 \text{ kW}$$

The turbine power is

$$P_{\text{opt}} = 1.38 \times 40^3 = 88.3 \text{ kW}$$

which is in agreement with the dc-link power. Based on the procedure for 40 rpm, the dc-link current–voltage characteristic over the speed range 30% to 100% is obtained as follows:

N (rpm)	V_d (V)	I_d (A)
15	231.5	20.1
20	308.5	35.8
25	385.2	56.0
30	462.2	80.6
35	538.5	109.9
40	614.0	143.6
45	688.6	182.7
50	761.6	226.6

2. The C_p-λ characteristic of a wind turbine is given by

$$C_p(\lambda) = C_{p,\text{opt}} \frac{\lambda}{\lambda_{\text{opt}}} \left(2 - \frac{\lambda}{\lambda_{\text{opt}}} \right)$$

in the working region of the characteristic. The static friction torque of the mechanical system is 10% of the torque developed by the machine at the rated wind speed and $\lambda = \lambda_{\text{opt}}$. The rotational frictional torque is proportional to the rotational speed of the turbine, and the frictional power loss at the rated speed is 2% of the maximum power captured by the turbine at the rated wind speed. The wind turbine can be coupled with any of the following variable-speed generating systems.

(i) A three-phase, four-pole induction machine with full-load slip $s_{\text{fl}} = 4\%$ is connected to the grid directly. The stator impedance can be neglected.

(ii) The same squirrel cage machine connected to the grid through a system similar to the one shown in Fig. 6.8.

(iii) A three-phase, four-pole salient-pole synchronous machine with reactances $X_q = 1$ p.u and $X_d = 1.6$ p.u and negligible stator resistance connected to the grid using the scheme shown in Fig. 6.9.

Each machine is rated such that it produces rated output power at the rated wind speed and the optimum tip speed ratio (λ_{opt}). Determine the following.

(a) The wind speed range (below the rated wind speed) over which each machine can operate.

(b) The relationship between the generated power and the wind speed in this wind speed range.

(c) For the third case, find the variation of the field current with the wind speed.

Solution

(i) Let the rated power output of the machine be P_{gr}. The mechanical power input to the machine under this condition is

$$P_{mr} = (1 + |s_{fl}|)P_{gr} \tag{6.38}$$

Since the frictional power loss is 2% of the power captured by the turbine under this condition,

$$P_{tr} = \frac{P_{mr}}{0.98} = 1.0204P_{mr}$$
$$= 1.02041(1 + |s_{fl}|)P_{gr}$$

The frictional power loss,

$$P_{lr} = 0.02P_{tr} = 0.02041P_{mr} = 0.02041(1 + |s_{fl}|)P_{gr} \tag{6.39}$$

The machine electrical speed under the rated condition is $\omega_{r,rated} = (1 + |s_{fl}|)\omega_e$. The Machine electrical speed at any other slip $|s|$ is

$$\omega_r = (1 + |s|)\omega_e$$

Now, the ratio of frictional power losses,

$$\frac{P_l}{P_{lr}} = \frac{\omega_r^2}{\omega_{r,rated}^2} = \frac{(1 + |s|)^2}{(1 + |s_{fl}|)^2} \tag{6.40}$$

where P_l is the frictional power loss at speed ω_r. Using Eqn (6.39) in Eqn (6.40) gives

$$P_l = \frac{(1+|s|)^2}{(1+|s_{fl}|)^2} P_{lr} = 0.02041(1+|s|)^2 \frac{P_{gr}}{(1+|s_{fl}|)}$$

If the wind speed corresponding to a slip $|s|$ is v, then

$$\frac{\lambda}{\lambda_{opt}} = \frac{\omega_r}{\omega_{r,rated}} \frac{V_R}{V} = \frac{1+|s|}{1+|s_{fl}|} \frac{V_R}{V} \tag{6.41}$$

$$\frac{P_t}{P_{tr}} = \frac{C_p(\lambda)}{C_p(\lambda_{opt})} \left(\frac{V}{V_R}\right)^3 = \frac{\lambda}{\lambda_{opt}} \left(2 - \frac{\lambda}{\lambda_{opt}}\right) \left(\frac{V}{V_R}\right)^3$$

The mechanical power input to the generator (turbine input minus the frictional power loss),

$$P_m = P_t - P_l$$

$$= \frac{\lambda}{\lambda_{opt}} \left(2 - \frac{\lambda}{\lambda_{opt}}\right) \left(\frac{V}{V_R}\right)^3 1.02041 (1 + |s_{fl}|P_{gr})$$

$$-0.02041(1+|s|)^2 \frac{P_{gr}}{1+|s_{fl}|} \tag{6.42}$$

Replacing λ/λ_{opt} in Eqn (6.42) by the expression in Eqn (6.41) we get

$$P_m = \frac{1+|s|}{1+|s_{fl}|} \left[2 - \frac{1+|s|}{1+|s_{fl}|} \frac{V_R}{V}\right] \left(\frac{V}{V_R}\right)^2 1.02041(1+|s_{fl}|)P_{gr}$$

$$-0.02041 \frac{(1+|s|)^2}{1+s_{fl}} P_{gr} \tag{6.43}$$

Assuming linear variation of the induction machine torque with slip,

$$\frac{T_e}{T_{er}} = \frac{|s|}{|s_{fl}|}$$

Therefore,

$$\frac{T_e \, \omega_r}{T_{er} \, \omega_{r,rated}} = \frac{P_m}{P_{mr}} = \frac{|s|}{|s_{fl}|} \frac{\omega_r}{\omega_{r,rated}}$$

or

$$\frac{P_m}{P_{mr}} = \frac{1+|s|}{1+|s_{fl}|} \frac{|s|}{|s_{fl}|}. \tag{6.44}$$

Using Eqn (6.38) in Eqn (6.44),

$$P_m = \frac{1+|s|}{1+|s_{fl}|}\frac{|s|}{|s_{fl}|}P_{mr}$$

$$= \frac{(1+|s|)|s|}{|s_{fl}|}P_{gr} \tag{6.45}$$

Combining Eqns (6.43) and (6.45) gives

$$\frac{1+|s|}{1+|s_{fl}|}\left[2 - \frac{1+|s|}{1+|s_{fl}|}\frac{V_R}{V}\right]\left(\frac{V}{V_R}\right)^2 1.02041(1+|s_{fl}|)P_{gr}$$

$$- 0.02041\frac{(1+|s|)^2}{1+s_{fl}}P_{gr}$$

$$= (1+|s|)\frac{|s|}{|s_{fl}|}P_{gr}$$

Suitable arrangement of the terms yields

$$2.04082\left(\frac{V}{V_R}\right)^2 - \frac{1+|s|}{1+|s_{fl}|}\left[1.02041\frac{V_R}{V} - 0.02041\right]$$

$$= \frac{|s|}{|s_{fl}|} \tag{6.46}$$

Setting $|s_{fl}| = 0.04$ in Eqn (6.46) and rearranging the terms, we get

$$|s| = \frac{2.08(V/V_R)^3 + 0.02V/V_R - 1}{1 + 25.46V/V_R} \tag{6.47}$$

At the minimum wind speed, $|s| = 0$. So

$$2.08\left(\frac{V}{V_R}\right)^3 + 0.02\left(\frac{V}{V_R}\right) - 1 = 0$$

the solution of which gives

$$\frac{V_{min}}{V_R} \approx 0.78$$

In the range $V_{min} \leq V \leq V_R$,

$$P_g = \frac{P_m}{1+|s|} = \frac{|s|}{|s_{fl}|}P_{gr} \qquad \text{[from Eqn (6.45)]} \tag{6.48}$$

Using Eqn (6.47) in Eqn (6.48) and substituting $|s_{fl}| = 0.04$, we get

$$\frac{P_g}{P_{gr}} = \frac{|s|}{|s_{fl}|}$$

$$= \frac{2.08 \, (V/V_R)^3 + 0.02 V/V_R - 1}{1.0184 V/V_R + 0.04}$$

The plot in Fig. 6.12(a) shows the variation of P_g/P_{gr} with V/V_R over the range 0.78 to 1.0.

(ii) In this case, the machine operates in the variable-voltage, variable-frequency mode. Therefore, the turbine can be made to operate at $\lambda = \lambda_{opt}$ for all $V_{min} \leq V \leq V_R$.

To find V_{min}, it should be noted that unlike in the case of a directly grid-connected machine (where we found out that $V_{min} \approx 78\%$ of V_R and $\lambda_{V_{min}} > \lambda_{opt}$), V_{min} is expected to be much lower than V_R. Moreover, with $\lambda = \lambda_{opt}$ the rotational speed of the turbine is much lower than its rated value. Therefore, the static frictional torque rather than the rotational frictional power loss determines the minimum wind velocity V_{min} (cut-in wind speed). At $V = V_{min}$, the static frictional torque equals the torque generated by the wind turbine at $\lambda = \lambda_{opt}$. With constant λ,

$$\frac{T_t}{T_{tr}} = \left(\frac{V}{V_R}\right)^2$$

At $V = V_{min}$, T_t (static frictional torque) $= 0.1 T_{tr}$ Therefore,

$$\left(\frac{V_{min}}{V_R}\right)^2 T_{tr} = 0.1 T_{tr} \tag{6.49}$$

i.e.,

$$V_{min} = 0.316 V_R$$

In the range $V_{min} \leq V \leq V_R$,

$$\lambda = \lambda_{opt}$$

Therefore,

$$P_t = \left(\frac{V}{V_R}\right)^3 P_{tr} = 1.02041 \left(\frac{V}{V_R}\right)^3 P_{mr} \qquad \text{[from Eqn (6.39)]}$$

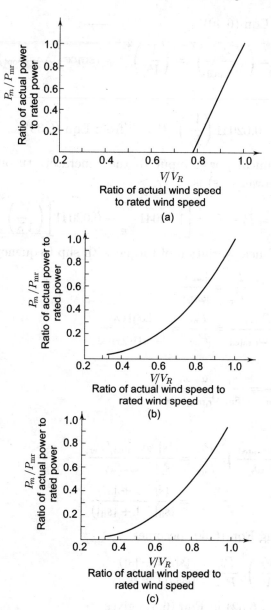

Fig. 6.12 Variation of generated power with wind speed: (a) case (i), (b) case (ii), (c) case (iii)

Now, from Eqn (6.40),

$$\frac{P_l}{P_{lr}} = \left(\frac{\omega_r}{\omega_{r,\text{rated}}}\right)^2 = \left(\frac{V}{V_R}\right)^2 \qquad (\text{since} \quad \lambda = \lambda_{\text{opt}}) \qquad (6.50)$$

So,

$$P_l = 0.02041 \left(\frac{V}{V_r}\right)^2 P_{\text{mr}} \qquad [\text{from Eqn } (6.39)]$$

The mechanical power input to the generator (turbine input – frictional power loss) is

$$P_m = P_t - P_l = \left[1.02041\frac{V}{V_R} - 0.02041\right]\left(\frac{V}{V_R}\right)^2 P_{\text{mr}} \quad (6.51)$$

Assuming linear variation of torque with slip frequency (ω_{sl}),

$$\frac{T_e}{T_{\text{er}}} = \frac{|\omega_{\text{sl}}|}{|\omega_{\text{slr}}|}$$

$$\frac{T_e\,\omega_r}{T_{\text{er}}\,\omega_{r,\text{rated}}} = \frac{P_m}{P_{\text{mr}}} = \frac{|\omega_{\text{sl}}|\,\omega_r}{|\omega_{\text{slr}}|\,\omega_{r,\text{rated}}}$$

or

$$\frac{P_m}{P_{\text{mr}}} = \frac{|s|}{|s_\text{fl}|}\frac{\omega_r\,\omega_e}{\omega_{r,\text{rated}}\,\omega_{\text{er}}}$$

Therefore,

$$\left(\frac{\omega_{r,\text{rated}}}{\omega_r}\right)^2 \frac{P_m}{P_{\text{mr}}} = \frac{|s|}{|s_\text{fl}|}\frac{\omega_{r,\text{rated}}/\omega_{\text{er}}}{\omega_r/\omega_e}$$

$$= \frac{|s|/(1+|s|)}{|s_\text{fl}|/(1+|s_\text{fl}|)}$$

which using Eqn (6.50) becomes

$$\left(\frac{V_R}{V}\right)^2 \frac{P_m}{P_{\text{mr}}} = \frac{|s|/(1+|s|)}{|s_\text{fl}|/(1+|s_\text{fl}|)} \qquad (6.52)$$

Using Eqn (6.52) in Eqn (6.51) gives

$$\frac{|s|}{1+|s|} = \left[1.02041\frac{V}{V_R} - 0.02041\right]\frac{|s_\text{fl}|}{1+|s_\text{fl}|}$$

For $|s_{fl}| = 0.04$,

$$\frac{|s|}{1+|s|} = \left[0.03925 \frac{V}{V_R} - 0.000785 \right]$$

i.e.,

$$\frac{1}{1+|s|} = 1.000785 - 0.03925 \frac{V}{V_R} \tag{6.53}$$

Now, for induction generator operation,

$$\frac{P_g}{P_{gr}} = \frac{1+|s_{fl}|}{1+|s|} \frac{P_m}{P_{mr}} \tag{6.54}$$

Using Eqns (6.51) and (6.53) in Eqn (6.54) and substituting $|s_{fl}| = 0.04$, we get

$$P_g = P_{gr} \left[1.04082 - 0.04082 \frac{V}{V_R} \right] \left[1.0204 \frac{V}{V_R} - 0.02041 \right] \left(\frac{V}{V_R} \right)^2$$

or

$$P_g = \left[1.063 \frac{V}{V_R} - 0.042 \left(\frac{V}{V_R} \right)^2 - 0.021 \right] \left(\frac{V}{V_R} \right)^2 P_{gr}$$

The plot in Fig. 6.12(b) shows the variation of P_g/P_{gr} with V/V_R over the range 0.316–1.0.

(iii) Since this is also a variable-speed generation system, the turbine can be made to operate at $\lambda = \lambda_{opt}$ for all $V_{min} \leq V \leq V_R$. By the same logic as in Eqn (6.49),

$$\left(\frac{V_{min}}{V_R} \right)^2 T_{tr} = 0.1 T_{tr}$$

or

$$V_{min} = 0.316 \, V_R$$

At any other wind speed V, $V_{min} < V \leq V_R$. Hence,

$$\frac{P_t}{P_{tr}} = \left(\frac{V}{V_R} \right)^3$$

$$\frac{P_l}{P_{lr}} = \left(\frac{\omega_r}{\omega_{r,\text{rated}}}\right)^2 \quad \text{[see Eqn (6.40)]}$$

$$= \left(\frac{V}{V_R}\right)^2 \quad \text{since } \lambda = \lambda_{\text{opt}}$$

Therefore,

$$P_m = P_t - P_l = \left(\frac{V}{V_R}\right)^3 P_{\text{tr}} - 0.02\left(\frac{V}{V_R}\right)^2 P_{\text{tr}} \quad (6.55)$$

Since $P_{lr} = 0.02 P_{\text{tr}}$,

$$P_{mr} = P_{\text{tr}} - P_{lr} = 0.98 P_{\text{tr}} \quad (6.56)$$

From Eqns (6.55) and (6.56),

$$\frac{P_m}{P_{mr}} = \left[1.02041\frac{V}{V_R} - 0.02041\right]\left(\frac{V}{V_R}\right)^2$$

Since the resistance of the machine is negligible,

$$\frac{P_m}{P_{mr}} = \frac{P_g}{P_{gr}}$$

Therefore,

$$P_g = \left[1.02041\frac{V}{V_R} - 0.02041\right]\left(\frac{V}{V_R}\right)^2 P_{gr} \quad (6.57)$$

The plot in Fig 6.12(c) shows the variation of P_g/P_{gr} with V/V_R over the range 0.316–1.0.

Let the machine phase voltage, phase current (fundamental component), and the rotational (electrical) speed reach their respective rated values V_{ar}, I_{ar}, and $\omega_{r,\text{rated}}$ at the rated wind speed and $\lambda = \lambda_{\text{opt}}$. Therefore,

$$P_{gr} = 3V_{\text{ar}}I_{\text{ar}} \quad (6.58)$$

Since the power factor is unity due to the input diode rectifier (vide Fig. 6.9),

$$I_a \propto I_{\text{dc}} \propto \omega_r \propto V$$

Therefore,

$$\frac{I_a}{I_{\text{ar}}} = \frac{V}{V_R} \quad (6.59)$$

and from Eqn (6.58), (6.59), and (6.57),

$$\frac{V_a}{V_{ar}} = \left(\frac{P_g}{P_{gr}}\right) \Big/ \left(\frac{I_a}{I_{ar}}\right)$$

$$= \left[1.02041\frac{V}{V_R} - 0.02041\right]\frac{V}{V_R} \qquad (6.60)$$

where V_a and I_a are the machine phase voltage and phase current (fundamental component), respectively, at any wind speed V and $\lambda = \lambda_{opt}$.

Let the induced voltage (per phase) be E_a under this condition. The relationship between E_a, V_a, and I_a is given by the phasor diagram shown in Fig. 6.13, in which X_q and X_d are the per-phase quadrature- and direct-axis synchronous reactances at the rated speed.

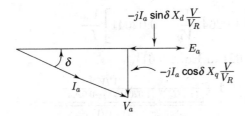

Fig. 6.13 The phasor diagram

From the phasor diagram,

$$V_a \sin\delta = I_a X_q \cos\delta \,\frac{V}{V_R}$$

$$V_a \cos\delta = E_a - I_a X_d \sin\delta \,\frac{V}{V_R}$$

Therefore,

$$\tan\delta = \frac{I_a X_q}{V_a}\frac{V}{V_R}$$

$$\cos\delta = \frac{1}{\sqrt{1 + \tan\delta^2}} = \frac{V_a V_R}{\sqrt{V_a^2 V_R^2 + I_a^2 X_q^2 V^2}}$$

and

$$\sin \delta = \sqrt{1 - \cos \delta^2} \doteq \frac{I_a X_q V}{\sqrt{V_a^2 V_R^2 + I_a^2 X_q^2 V^2}}$$

$$\begin{aligned}
E_a &= V_a \cos \delta + I_a X_d \sin \delta \frac{V}{V_R} \\
&= \frac{V_a^2 V_R^2 + I_a^2 V^2 X_q X_d}{V_R \sqrt{V_a^2 V_R^2 + I_a^2 X_q^2 V^2}} \\
&= V_a \frac{1 + \frac{X_q X_d}{(V_a/I_a)^2} (V/V_R)^2}{\sqrt{1 + \frac{X_q^2}{(V_a/I_a)^2} (V/V_R)^2}}
\end{aligned}$$
(6.61)

From Eqns (6.59) and (6.60),

$$\frac{V_a}{I_a} = \left[1.02041 \frac{V}{V_R} - 0.02041 \right] \frac{V_{\mathrm{ar}}}{I_{\mathrm{ar}}}$$
(6.62)

Using Eqn (6.62) in Eqn (6.61),

$$E_a = V_a \frac{1 + \frac{X_q X_d}{(V_{\mathrm{ar}}/I_{\mathrm{ar}})^2} \frac{(V/V_r)^2}{(1.0204 V/V_r - 0.02041)^2}}{\sqrt{1 + \frac{X_q^2}{(V_{\mathrm{ar}}/I_{\mathrm{ar}})^2} \frac{(V/V_R)^2}{(1.02041 V/V_R - 0.02041)^2}}}$$

But

$$\frac{X_q}{V_{\mathrm{ar}}/I_{\mathrm{ar}}} = \bar{X}_q = \text{per-unit } q\text{-axis reactance at rated speed}$$

$$\frac{X_d}{V_{\mathrm{ar}}/I_{\mathrm{ar}}} = \bar{X}_d = \text{per-unit } d\text{-axis reactance at rated speed}$$

So,

$$E_a = V_a \frac{1 + \bar{X}_q \bar{X}_d \frac{(V/V_R)^2}{(1.02041 V/V_R - 0.02041)^2}}{\sqrt{1 + \bar{X}_q^{\,2} \frac{(V/V_R)^2}{(1.2041 V/V_R - 0.0241)^2}}}$$

Substituting the values, we get

$$E_a = V_a \frac{2.64 (V/V_R)^2 - 0.042 V/V_R + 0.00042}{(1.02041 V/V_R - 0.02041) \sqrt{2.04 (V/V_R)^2 - 0.042 V/V_R + 0.00042}}$$
(6.63)

Now $E_a \propto I_f \omega_r$. At $I_f = I_{f,\text{rated}}$ and $\omega_r = \omega_{r,\text{rated}}$, $E_{ar} = 1.838 V_{ar}$.

$$\frac{E_a}{E_{ar}} = \frac{I_f}{I_{f,\text{rated}}} \frac{\omega_r}{\omega_{r,\text{rated}}} = \frac{I_f}{I_{f,\text{rated}}} \frac{V}{V_R}$$

Therefore,

$$\frac{I_f}{I_{f,\text{rated}}} = \frac{V_R}{V} \frac{E_a}{E_{ar}} \times \frac{V_{ar}}{V_{ar}} = \frac{V_R}{V} \frac{E_a}{1.838 V_{ar}} \tag{6.64}$$

Using Eqn (6.60) in Eqn (6.64) gives

$$\frac{I_f}{I_{f,\text{rated}}} = \left(1.02041 \frac{V}{V_R} - 0.02041 \right) \frac{E_a}{1.838 V_a} \tag{6.65}$$

Substitution of Eqn (6.63) in Eqn (6.65) gives

$$\frac{I_f}{I_{f,\text{rated}}} = \frac{2.64(V/V_R)^2 - 0.042 V/V_R + 0.00042}{1.838 \sqrt{2.04(V/V_R)^2 - 0.042 V/V_R + 0.00042}}$$

$$= \frac{1.85}{1.838} \times \frac{(V/V_R)^2 - 0.016 V/V_R + 0.00016}{\sqrt{(V/V_R)^2 - 0.02 V/V_R + 0.0002}}$$

which verifies that

$$\frac{I_f}{I_{f,\text{rated}}} \approx \frac{V}{V_R} \quad \text{[vide Eqn (6.23)]}$$

In part (ii) as well as part (iii), we have assumed that at $V = V_{\min}$ and $\lambda = \lambda_{\text{opt}}$, the torque generated in the turbine just equals the static frictional torque. Strictly speaking, for self-starting, the starting torque at $V = V_{\min}$ should have been considered. However, fast horizontal-axis wind turbines have very low starting torques and may not be capable of self-starting even at the rated wind speed. Therefore, in a grid-connected system, the electrical machine is operated as a motor (or, some other starting method is applied) to accelerate the turbine up to $\lambda = \lambda_{\text{opt}}$. Thereon, the turbine should be capable of sustaining its rotation at the cut-in speed. Therefore, the torque generated by the turbine at $\lambda = \lambda_{\text{opt}}$ has been used in the problems.

Summary

In order to extract the maximum amount of power from wind, the tip speed ratio must be kept fixed at the optimal value while the wind speed fluctuates continuously. Consequently, the electrical generator should be able to operate at variable speed. Since variable-speed generation is most energy efficient and has high promise in future applications, this chapter is exclusively devoted to the variable-speed-based generation schemes.

Even though constant-speed wind turbines with grid-connected squirrel cage induction generators dominate the wind market today, there is a clear trend over the past decade towards the increasing use of variable-speed wind turbines. The advantages of variable-speed operation (as opposed to constant-speed operation) are (a) the ability to extract more power, (b) control of reactive power, (c) reduced flicker in the power output, (d) less audible noise, (e) lower mechanical load, etc. Variable-speed operation can be carried out with or without a gear box (direct drive). The double-output induction generator uses a gear box whereas a wound-field or permanet magnet rotor synchronous generator generally uses direct drive technology. Low maintenance due to the absence of a gear box, higher efficiency at low wind speeds, etc. are some of the advantages compared to the schemes that use a gear box. The disadvantages are a large diameter and a larger electronic converter (100% of the rated power compared to about one-third of the power with a doubly fed induction generator).

A variable-speed synchronous generator with a gear box is lighter and cheaper than the direct drive machine, but with respect to converter size it has the same deficiency, i.e., a larger, more expensive converter (100% rated power). The variable-speed squirrel cage induction generator has the advantage of using cheaper generators but suffers from the same demerit with respect to the converter size and the losses in it. So far, the maximum rating for variable-speed direct drive synchronous generators combined with pitch control, has reached 1.5 MW, against 4.5 MW obtainable with variable-speed doubly fed induction generators.

This chapter discusses the benefits of and the requirement for variable-speed operation of turbines. Various schemes using the cage rotor induction machine as the variable-speed generator, with a forced-commutated rectifier and a PWM inverter/rectifier for processing variable-voltage, variable-frequency power are discussed. For direct drive applications, synchronous machines of the wound rotor as well as the permanent magnet types are attractive.

The principle of power generation control of synchronous machines is studied, and various control schemes are presented.

Problems

1. The variation of the torque coefficient C_T over the operating range of a prototype wind turbine of diameter 3 m is $0.175 - 0.025\lambda$, where λ is the tip speed ratio. The turbine drives a permanent magnet dc generator which feeds a resistive load of 25 Ω. The emf constant of the generator is 0.25 V/rpm. Find the speed of the aerogenerator if the wind speed is 11 m/s and the system efficiency is 75%. Take the density of air as 1.293 kg/m^2 and ignore the armature resistance of the generator.

2. The per-phase reactance and resistance of a three-phase generator are 2.75 Ω and 0.875 Ω, respectively. The open-circuit line-to-line voltage at the operating frequency is 400 V. It is connected to a dc line through a diode-bridge rectifier. If the generated power is 18 kW, what should the dc-link voltage and current be? Assume level dc-link current and voltage.

7

Hybrid Energy Systems

Many sparsely populated remote areas, where the demand for power is low, need a stand-alone electrical power source, as it is highly uneconomical to extend power distribution lines to them. Renewable energy sources such as wind or solar energy can be utilized as independent sources of electrical power in such areas. But the nature of these sources is very different from that of conventional ones. The supply from such sources depends heavily on weather conditions and is intermittent and/or fluctuating. Besides this, the variation in supply may not match the distribution of the demand. Because of such problems, it may not be possible for these renewable energy systems to provide a continuous supply over long periods of time. To achieve reliability of supply, system designers consider measures such as combination with rechargeable batteries for energy supply during peak load periods and operation with some other type of generator. Batteries for continuous energy supply are expensive; they also require periodic charging, for which a separate source of power is required.

Diesel-driven alternators provide a reliable continuous source of electrical energy. They have attained a high level of technology, and are highly dependable, if maintained properly. Although the initial cost of diesel generators may be low compared to renewable generators of compatible size, they are characterized by high running cost, poor fuel efficiency, high transportation cost, and relatively high cost of maintenance and operation in remote areas. These disadvantages can be attributed to the operation of the diesel unit continuously at low load factor, which is unavoidable as the diesel set must accommodate the peak load. It is not uncommon for the designer to set the capacity of the diesel generator at a level higher than the peak load so as to provide a safety margin for a sudden demand due to additional load and also leave scope for future extensions. Moreover, load fluctuations through a single day and between one day and another are very common. As the load on the diesel generator falls, its efficiency goes down, and at zero load the engine burns 20–30% of the fuel required for rated load. All these factors make diesel generation expensive.

However, some efficient systems can be contrived by integrating renewable energy (wind/solar) systems and battery inverter subsystems into diesel generator sets. Such *hybrid systems* can have a number of benefits. Under favourable wind conditions, wind turbines can partially relieve the diesel set of its load, thereby saving some fuel. If sufficient wind power is available, the diesel set can be shut down and power demands met by the wind generator, thus offsetting the high cost of running the diesel generator. In the case of a fall in the wind speed, the diesel generator can be started in order to cope with the load above that supplied by the wind generator.

Photovoltaic power supplies can also be employed to form an integrated hybrid system with the wind generator. In a thermally driven wind area, strong winds occur towards or after sunset, and such winds are often associated with overcast conditions. Thus, owing to the complementary nature of solar and wind energy, a hybrid solution using turbines and photovoltaic modules may provide a reliable configuration over a long period. In a wind–solar hybrid system, wind power plays a dominant role, because photovoltaic production is costlier.

7.1 Diesel Generator and Photovoltaic System

7.1.1 Diesel Engine

A diesel-fuelled power plant is a form of internal combustion engine which uses high-speed diesel fuels, works on the compression–

ignition principle, and employs a four-stroke cycle of operation. The four strokes are the suction stroke, compression stroke, expansion (power) stroke, and the exhaust stroke. During the suction stroke, the piston of the engine cylinder moves towards the crank end of the cylinder (downward movement) and fresh air is sucked into the cylinder through an inlet valve. In the compression stroke, the piston moves towards the cover end of the cylinder (upward movement) and compresses the air to the extent of raising its temperature to well above 558 °C. At the end of this stroke a fine spray of fuel is injected into the cylinder by an injector nozzle, resulting in the spontaneous ignition of the fuel–air mixture. The hot gas inside the cylinder develops a high pressure, which pushes the piston downward, delivering power to the shaft. This is the power stroke. In the final stroke (i.e., the exhaust stroke), the piston moves upward and the burnt gas escapes through an exhaust valve. This completes one cycle, which is repeated.

As the distance travelled by the piston is fixed, the volume of air drawn is always the same. Hence, the work done every two revolutions (four strokes) depends on the amount of fuel injected. Fuel injection is controlled by the opening of a valve called the throttle valve. For a set throttle opening, the ideal torque T_d is independent of the speed over its working speed range and proportional to the fuel consumption m_d per stroke cycle, as shown below. From the power balance condition,

$$T_d \omega_m = \text{stroke cycles per second} \times \text{fuel injection per}$$
$$\text{stroke cycle} \times \text{heat value of the fuel}$$
$$= (\omega_m/4\pi)m_d H$$

i.e.,

$$T_d = m_d H/4\pi$$

However, owing to losses, T_d shows a drooping trend with increase in speed. Figure 7.1 shows the general nature of the speed–torque curves of a diesel engine. It also shows how the characteristics shift for an increased throttle opening, i.e., for an increased fuel–air ratio. A diesel engine has no starting torque; it is started by some auxiliary means. It has a small overload capacity.

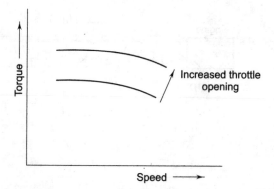

Fig. 7.1 Torque–speed relationship of a diesel engine

In order to investigate the dynamic behaviour of an autonomous system supplied by diesel and other power sources, an appropriate dynamic model of the diesel engine with governor is required along with models of the other units. The block diagram shown in Fig. 7.2 presents an adequate model of the diesel engine with a speed governing system. The *speed regulation parameter* R in the feedback path represents the characteristic of the governor which, for a constant setting, makes an increment in the diesel engine static power output proportional to the static frequency drop. The unit of R is Hz/kW. An integral control with gain K_I is introduced to eliminate the static frequency error. The model utilizes a first-order representation of the governor with a time constant τ_G and the fuel actuator gain constant K_G. The governor controls the engine fuel consumption rate $\dot{m}_d$, which in turn determines the developed torque of the engine. The model also includes a time delay τ and the combustion efficiency ε, which is the ratio of the power developed by the engine to the heat consumed during the same period. As the ideal torque is proportional to m_d, the actual torque developed by the engine, considering the efficiency, will be

$$T_{\text{dm}}(s) = (C\epsilon e^{-s\tau})\dot{m}_d$$

where C is an appropriate constant.

The feedback constant C_f accounts for the mechanical loss torque T_{df}, assumed proportional to the speed of the engine. The effective torque T_{de} of the engine is then $T_{\text{dm}} - T_{\text{df}}$. If T_{el} is the electrical load torque, the mechanical response is set by the torque

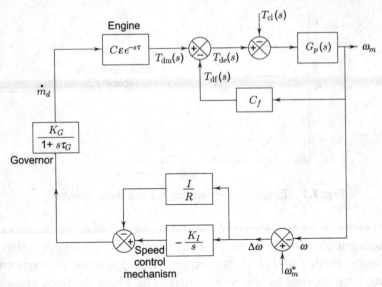

Fig. 7.2 Block diagram of a diesel engine

difference $T_{de} - T_{el}$. The inertia and the damping constants of the unit are represented by the transfer function $G_p(s)$.

7.1.2 Photovoltaic Power Generation

Solar cells, modules, and arrays

Photovoltaic (PV) power generation involves the technology of converting thermal radiation from the sun directly into electrical energy. Such conversion takes place through units called solar cells, which are made of semiconductor materials with differently doped monocrystalline silicon layers forming a *pn* junction. A solar cell is therefore nothing but a diode with a large surface area and a large area of the *pn* junction. At this junction, a potential barrier is developed, which prevents the majority carriers (electrons in the *n*-type material and holes in the *p*-type material) from crossing the junction. When incident solar radiation (called *insolation*) is absorbed by the semiconductor material, electron–hole pairs are created by the energy of the photons. The minority carriers (electrons in the *p*-region and holes in the *n*-region) diffuse to the junction, cross it, and lower the barrier potential. A voltage equal to the decrease in the barrier potential appears across the metallic contacts of the diode, and current flows when a load is connected.

A basic cell configuration is shown in Fig. 7.3. The load is connected to the cell through the metallic contacts. The metallic contacts on the top (illuminated) surface form a conducting mesh through which radiation is allowed to fall on the diode material. The opposite surface (dark side) is provided with a conducting metallic base. To enhance the absorption of radiation by reducing reflection, an antireflection coating of SiO_2 is applied on the top surface. For mechanical protection, the cell is placed under a glass cover and attached to it by a transparent adhesive.

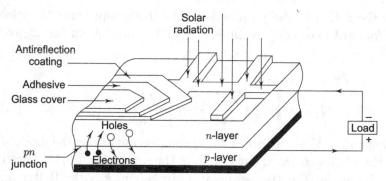

Fig. 7.3 Basic constructional features of a PV cell

A single silicon solar cell, with efficiency values between 15% and 18%, develops an open-circuit voltage of the order of 0.65 V and generates a maximum current density between 35 and 40 mA/cm^2. Thus for a cell with an area of 75 cm^2, the power developed is limited to about 1.5 W. In a photovoltaic system, single solar cells are combined in a series–parallel arrangement to form a module, and several modules are interconnected in series–parallel to make a solar array. In order to ensure the desired voltage and current ratings, a suitable choice of series and parallel connections of arrays is made.

The equivalent circuit

Figure 7.4 shows the circuit model of a photovoltaic (PV) generator. The behaviour of this circuit determines the output characteristics of a PV generator. In this equivalent circuit, R_s represents the internal (series) resistance of the system, and R_{sh} represents the effect of current leakage to the ground and the recombination of the electrons and holes before they are separated

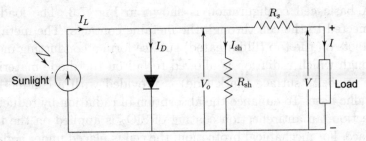

Fig. 7.4 Equivalent circuit of a PV module

by the action of the pn junction. The diode represents the effect of forward biasing of the pn junction. The output current is given by

$$I = I_L - I_D - I_{\text{sh}}$$

$$= I_L - I_o \left[\exp \left(\frac{qV_o}{nkT} \right) - 1 \right] - \frac{V_o}{R_{\text{sh}}} \tag{7.1}$$

where $V_o = V + IR_s$ is the voltage appearing across the diode, I_L is the photogenerated current, I_o is the reverse saturation current of the diode, T is the temperature in kelvin, k is the Boltzmann constant (1.38×10^{-23} J/K), n is the ideality factor (≈ 1.92), and q is the electronic charge (1.6×10^{-19} C).

When the cell is open-circuited, i.e., when $I = 0$, the voltage V_o across the shunt branch appears as the open-circuit voltage V_{oc} across the output terminals. Neglecting the effect of R_{sh} in Eqn (7.1), the open-circuit terminal voltage becomes

$$V_{\text{oc}} = \frac{nkT}{q} \ln \left(\frac{I_L}{I_o} + 1 \right)$$

The output characteristics

The output characteristics of a photovoltaic generator are described by the current–voltage (I-V) and power–voltage (P-V) curves shown in Fig. 7.5 with the insolation and the temperature as parameters. The output power is the product of the terminal voltage and the terminal current. For low output voltage, a PV generator behaves like a constant-current source, and for low output current, like a constant-voltage source. The source current varies almost linearly with insolation, whereas the open-circuit voltage is logarithmically related to the insolation, and

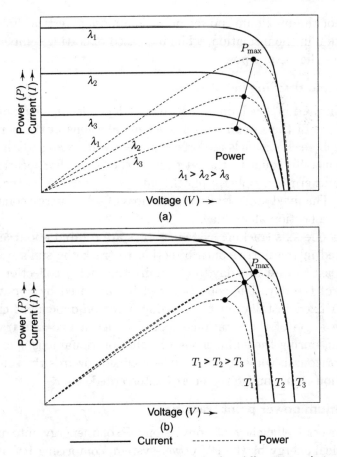

Fig. 7.5 *I-V* characteristics of a PV module for different insolation and temperature levels

hence shows little fluctuation with change in insolation. With rise in temperature, the open-circuit voltage reduces, but the short-circuit current increases. However, the per-unit decrease in the open-circuit voltage is much more than the increase in the short-circuit current. The maximum power points (MPP) occur at voltages around the knee points in the *I-V* characteristics. The maximum power obtainable from a panel is given by the area of the largest rectangle that can be drawn under the *I-V* curve. The *fill factor* is defined as the ratio of the maximum sized rectangle under the *I-V* curve to the product of the open-circuit voltage and the short-circuit current. Note that at constant temperature, the locus of the maximum power point is almost a vertical line with variation in the insolation, while increased operating temperature reduces the output power.

Fixed and tracking installations

The insolation level on a PV panel depends on the angle of incidence of the sun's rays. In fixed installations in the northern hemisphere, PV panels are oriented east–west, facing south at an angle usually equal to the latitude of the place. For locations in the southern hemisphere, panels have to face north at the same angle. The fixed angle leads to some loss of solar energy compared to the collection at midday.

In a one-axis tracking system, arrays are rotated about an axis oriented in the north–south direction to track the sun's position from east to west in a day to maximize the energy collection at all times of the day. The collection can be increased by about 35% over a fixed installation by allowing a season-dependent change in the angle of the rotational axis. In such a two-axis tracking system, during the summer months the rotational axis is close to the horizontal, while in winter it is inclined towards the southern direction to maximize the energy intercepted.

Maximum power point tracking

The overall efficiency of conversion of solar energy into usable electrical energy by the PV power system, comprising PV arrays, converters, cable connections, etc., is quite low ($< 6\%$). Because of the specific nature of the I-V characteristic, the output power is maximized at a specific load for a given level of solar insolation and cell temperature. Moreover, unlike a conventional power generating system, where the fuel input can be controlled depending on the power demand, the input to the PV system is not controllable. The PV array in the system accounts for more than 50% of the total cost. Under these circumstances, it makes good economic sense to operate the solar array in a such way as to extract maximum power for any insolation level and operating temperature.

In order to provide reliable supply from stand-alone generating systems using renewable energy sources, it is necessary to provide battery backup. If the extracted power during daytime is higher than the demand, the balance is used to charge batteries, which are in turn used to meet the demand when solar power is insufficient or unavailable. When the batteries are fully charged, the extra power is disposed of into dummy loads.

The output voltage of the solar array is generally not the same as the voltage of the dc-link connected to the battery, which operates at an almost constant voltage. Self-adaptive dc–dc converters convert the photovoltaic panel output voltage into the dc-link voltage, as required by the battery or the load. So a change in the converter's duty cycle alters the input voltage to the converter, which is also the panel's output voltage. The controller, through its adaptive action, can then adjust the input voltage to be equal to the panel's maximum power point voltage. Figure 7.6 shows a typical stand-alone PV power system with an integrated maximum power point tracking (MPPT) converter and battery backup.

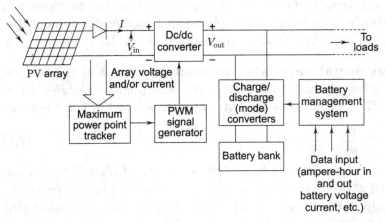

Fig. 7.6 Stand-alone PV power system with an MPPT converter and battery backup

Control algorithms for MPPT can be based on the techniques given below. Their merits and limitations are judged in terms of simplicity, accuracy, adaptability to temperature and irradiance variations, control circuit complexity, and relative implementation cost.

Control system with current or voltage reference In this method of control, the actual PV array output voltage (or current) is compared with a reference signal, which corresponds to the maximum power point. The error is processed through a proportional-integral (PI) controller, the output of which is used to control the duty cycle of the PWM control signal for the dc–dc converter. Various methods exist to deduce the reference voltage and current, which are either constant or derived from the change of the output power of the PV array.

The maximum power point voltage is always close to a fixed percentage, generally 78% of the open-circuit voltage with a tolerance band of ±2%, while the maximum power point current is close to 90% of the short-circuit current. These values can be used as references for the control loop. In another current-controlled maximum power point tracking system, the current reference is derived from the comparison of the PV array output power before and after a change in the duty cycle of the dc–dc converter. The reference voltage in a voltage-controlled system is calculated from the difference between the output powers at the previous and present voltage levels. For implementing such feedback loops, the voltage and current values are sampled at regular intervals.

Incremental impedance method This method is based on the principle that at the maximum power point on the P-V curve $dP/dV = 0$, and since the output power is $P = VI$, it follows that

$$\frac{dI}{dV} + \frac{I}{V} = 0 \tag{7.2}$$

The left-hand side of Eqn (7.2) is calculated digitally and compared with a zero reference. The error is applied to a PI controller to adjust the duty cycle of the dc–dc converter to a value that would satisfy the condition in Eqn (7.2).

Operation around the peak power point It is very difficult to hold the PV array output power precisely at a voltage where dP/dV or dP/dI equals zero. In another method, the duty ratio is changed in one direction by a preset value and the PV array output voltage and current are sampled at equal time intervals. Let the powers measured at the $(n-1)$th and nth sampling be $p(n-1)$ and $p(n)$. If $p(n) - p(n-1)$ shows a positive sign, the duty cycle is changed in the same direction by a definite step in order to increase the output power. If the power difference in the two consecutive observation instants yields a negative sign, the duty cycle is changed by the same step in the reverse direction so as to increase the output power. In this way, depending upon the step change in the duty cycle, dP/dV can be kept close to zero within a preset band.

7.2 Wind–Diesel Hybrid System

7.2.1 System With No Storage

Unfortunately, wind–diesel integration is not as simple and straightforward an approach as one may think. Many problems may arise during this process. When wind generators are tied with the utility grid, the system voltage and frequency are virtually constant and independent of wind energy variations. However, in the case of small autonomous networks integrating wind and diesel generators, wind power fluctuations significantly affect the system voltage and frequency, subject to the control policies and performance of the diesel sets connected to the network.

With high wind energy output, the average diesel load is low. With the reduction in load, the specific energy consumption increases and hence the efficiency of the diesel generator goes down. With no load, a diesel generator consumes about 30% of the amount of fuel required at full load. Prolonged running at low load shortens its working life. Many manufacturers specify a minimum operating load, usually of the order of 40% of the full load. For this reason the diesel generator is shut down when the wind power is sufficient to meet the load demand. At times of high wind generation, a mismatch between the wind turbine output and the load may lead to stability problems and wind energy wastage. Arrangements can be made for the energy generated in excess of the electrical load to be taken care of by diverting it for space and water heating, refrigeration, or irrigation pumping, thereby avoiding wastage.

It should be noted that storing energy in a suitable medium and reconverting that stored energy into normal ac voltage involves both cost and complexity. The use of 'dump load' can prevent the wind turbine from driving the diesel engine, and when the diesel engine is driven by the wind turbine, frequency control is lost.

Dump load can also be used to maintain minimum load on the diesel generator. However, this is a wasteful method.

In the case of a sudden drop in the wind speed and/or an increase in the load demand, the diesel engine might need to be restarted. Fluctuations in the wind speed and the load demand may result in a larger number of stop/start cycles of the diesel generator. Such frequent starting of the diesel machine causes fuel wastage and wears out the engine. Moreover, starting the diesel generator after a drop in the wind speed takes some time, causing discontinuities in the supply. To overcome these operational problems, the use of multiple diesel units is suggested, where one unit is run continuously, meeting the minimum load requirement, and others are started as and when required.

The wind–diesel hybrid system is suitable for small wind energy penetration and small installations, when the wind turbine can operate in parallel with diesel sets, reducing the average diesel load, thus saving fuel.

Figure 7.7 illustrates the configuration of a typical diesel–wind hybrid power network, which includes a diesel-driven synchronous generator, a controlled-rectifier-driven dump resistive load, a wind turbine generator, etc. A unique system design cannot be followed in all situations. Approaches to system design depend on consumer demand, the possibility of demand control, the availability of wind energy and its variability, the existing generators in operation, the local distribution system, and the likely economic benefits, coupled with reliability of supply.

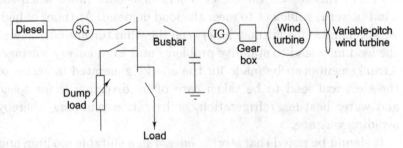

Fig. 7.7 Single-line diagram of a wind–diesel hybrid power system

7.2.2 System With Battery Backup

Operational problems with wind–diesel systems can be overcome to a significant extent with the inclusion of energy storage elements. The battery as a storage device has the advantage that it is easily amenable to rapid changes in energy flow in either direction without undue wear and maintenance problems. With diesel generation shut down and wind speed falling suddenly, a battery can meet the power demand before the diesel generator is started again. An immediate question about the capacity of the bank arises. This capacity should be sufficient to cover the diesel startup time. Second, frequent diesel startups result in fuel wastage, decreased efficiency, and frequent maintenance requirements. Therefore, the storage capacity of the battery bank should be sufficient to cover the short-time fluctuations in wind speed (high-frequency turbulence) with periods varying from a few seconds to a few minutes. It follows, therefore, that the inclusion of a battery with a low storage capacity, which can supply energy for a few minutes, will have a significant impact on the required number of diesel start/stop cycles.

A small battery capacity in the case of high-frequency turbulence results in an increase in the charge/discharge rates and cycles, which may severely shorten the life of the battery. So these parameters must be properly taken care of in the system design and control strategy. Merely selecting a small storage capacity to meet the load deficiency will not suffice. Moreover, an increase in battery capacity saves diesel fuel. But beyond a certain point, additional capital cost for increased storage may not be justified.

Another advantage of battery backup lies in its ability to improve wind energy utilization. During low wind, the battery can meet the load demand, and when wind speed is sufficient, the excess wind energy can be used to recharge the battery, thereby saving diesel fuel. Marginal quantities of load above the rated load of the wind generator can be transferred to the battery instead of running the diesel generator inefficiently at low load. Likewise, during times of low demand and low wind, when the diesel unit is brought into service, the battery bank can provide an additional dump load to the diesel generator, enabling it to maintain a minimum generation.

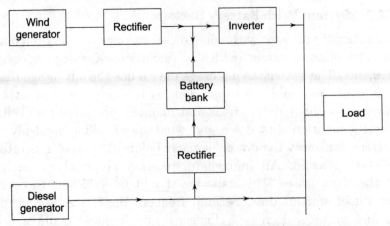

Fig. 7.8 Block diagram of a wind–diesel–battery hybrid system

A variety of hybrid systems combining diesel generators with battery inverter subsystems and incorporating wind generators is possible. The system design and the control of operation depend on factors such as consumer requirement, site location, and system economics. The block diagram in Fig. 7.8 schematically shows a hybrid system in which, during high wind, the output from the variable-speed wind generator is rectified and then inverted for ac output at normal voltage and frequency. The excess power is used to charge the battery. In this system, a bidirectional inverter can act in the reverse direction to charge the battery. Variable-speed operation allows maximum energy capture. It is normal practice to use a synchronous generator for the diesel system, and an induction or synchronous generator for the wind turbine. The system provides very flexible operation. If so desired, one of the generators can be shut down for efficient operation of the other unit. During low winds, the wind generator along with the battery can meet the load. If the consumer demand cannot be met by the wind generator and the battery, the diesel generator can be started, and the battery may be withdrawn for adequate loading of the diesel generator. The several advantages of this configuration are as follows.

(a) Low-load operation and a large number of start/stop cycles of the diesel engine can be avoided, minimizing the diesel maintenance cost and achieving better fuel efficiency.

(b) Optimum scheduling of load distribution between the generators and the battery is possible.

(c) None of the units needs to be individually sized to meet the peak demand.

7.3 Wind–Photovoltaic Systems

In many regions of the world, wind generators and photovoltaic cells are combined to provide year-round renewable energy to non-grid-connected households. This is possible since variations in wind and solar power resources are usually complementary. A wind generator is thus an excellent supplement to the PV system and vice versa. Moreover, interfacing of wind generators and PV cells minimizes the battery capacity and extends the battery bank life compared to the storage requirement in solar- or wind-only systems. Figure 7.9 shows a typical wind/PV hybrid system configuration.

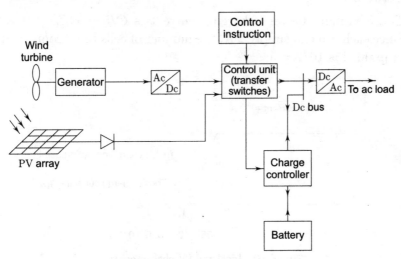

Fig. 7.9 Configuration of a typical wind–PV system

The ac output of the wind generator feeds a rectifier which is connected in parallel to the PV array through a controller to a dc bus. The dc bus also serves as a connection point for the battery through a charge controller. The blocking diode protects the PV array from voltage spikes and prevents the flow of current in the reverse direction at low irradiation. The controller decides the connection of the generating system/battery supply, or its charging, in specific situations and requirements.

Worked Examples

1. A 12-V, 50 Ah capacity battery is to be charged directly using a solar photovoltaic panel. Each cell in the panel has an open-circuit voltage of 0.65 V, and can supply a current of 2 A at maximum illumination. The cell voltage drops by 10% from its open-circuit value before it enters the constant-current mode. The battery bulk charge rate should not exceed $C/5$ and the trickle charge rate should not be more than 10% of the maximum bulk charge rate. The battery overcharges by 5% (from its nominal voltage) during trickle charge. Find the number of cells to be connected in series and parallel.

Solution

The maximum battery charging current is $C/5 = 50/5 = 10$ A. Since each cell can supply 2 A, the number of cells to be connected in parallel is $10/2 = 5$.

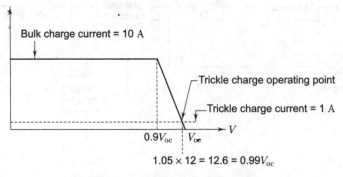

Fig. 7.10 Idealized PV characteristic

To find the number of cells to be connected in series, let us consider the idealized PV characteristics shown in Fig. 7.10. From the figure, at the trickle charge operating point, the panel output voltage should be 12.6 V. But from the given data,

$$0.99V_{oc} = 12.6 \text{ V}$$

i.e.,

$$V_{oc} = 12.73 \text{ V}$$

So the number of cells to be connected in series $= 12.73/0.65 \approx 20$.

2. Figure 7.11 shows the idealized load variation of a small isolated power system over a day. The system is supplied by a wind–diesel hybrid. The diesel generator is designed to supply 100% peak load. It consumes 30% of the rated fuel intake at no load. The fuel intake increases linearly with load to 100% thereafter. Assuming favourable wind conditions, find the optimum rating of the wind generator so that the cost of energy is optimized.

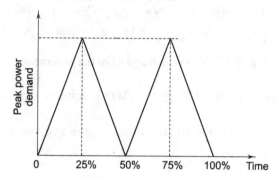

Fig. 7.11 Idealized load variation of an isolated system

Solution

In this context, a favourable wind condition would mean that the wind turbine is capable of generating power according to the load demand up to its rated capacity. If the capacity of the wind turbine is $x\%$ of the peak load demand, the diagram shown in Fig. 7.12 can be drawn. The diesel generation P_D can be expressed as

$$0 \le t \le \frac{x}{4}\%, \quad P_D = 0$$

$$\frac{x}{4}\% \le t \le 25\%, \quad P_D = (4t - x)$$

$$25\% \le t \le \left(50 - \frac{x}{4}\right)\%, \quad P_D = (200 - x - 4t)$$

$$\left(50 - \frac{x}{4}\right)\% \le t \le 50\%, \quad P_D = 0 \tag{7.3}$$

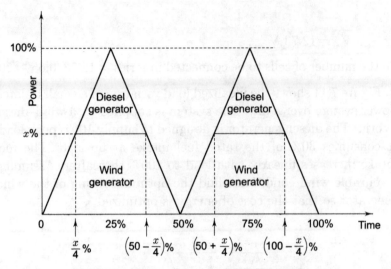

Fig. 7.12 Power demand from the generators

For the time interval $50\% \leq t \leq 100\%$, similar expressions can be written.

The fuel consumption of the diesel engine is given by

$$f = 30 + 0.7 P_D \tag{7.4}$$

The diesel engine is turned off when $P_D = 0$. Therefore, the fuel consumption as a function of time is given by

$$0 \leq t \leq \frac{x}{4}\%, \quad f = 0$$

$$\frac{x}{4}\% \leq t \leq 25\%, \quad f = [30 + 0.7(4t - x)]$$

$$25\% \leq t \leq \left(50 - \frac{x}{4}\right)\%, \quad f = [30 + 0.7(200 - x - 4t)]$$

$$\left(50 - \frac{x}{4}\right)\% \leq t \leq 50\%, \quad f = 0 \tag{7.5}$$

Therefore, the daily fuel consumption is

$$f_D = 2 \int_{(x/4)\%}^{25\%} (30 + 2.8t - 0.7x) dt$$

$$+ 2 \int_{25\%}^{(50 - x/4)\%} (170 - 2.8t - 0.7x)\, dt \tag{7.6}$$

Carrying out the integration gives

$$f_D = \frac{1}{2}(100 - x)(130 - 0.7x) \tag{7.7}$$

Let the cost of fuel for the diesel engine be C_d when run at 100% load all the time. Further, let the daily return on the investment on the wind turbine with 100% capacity be C_w and on the diesel generator be C_f. The daily energy cost is

$$C = \frac{C_d}{10^4} f_D + \frac{C_w}{100} x + C_f \tag{7.8}$$

The use of Eqn (7.7) in Eqn (7.8) gives

$$C = \frac{(100 - x)(130 - 0.7x)C_d}{2 \times 10^4} + \frac{C_w x}{100} + C_f \tag{7.9}$$

i.e.,

$$2 \times 10^4 C = C_d(100 - x)(130 - 0.7x) + 200 C_w x$$
$$+ 2 \times 10^4 C_f \tag{7.10}$$

For optimum wind turbine rating,

$$\frac{d}{dx}(2 \times 10^4 C) = 0 \tag{7.11}$$

or

$$200 C_w + C_d[(0.7x - 130) + 0.7(x - 100)] = 0 \tag{7.12}$$

or

$$x = \frac{100}{0.7}\left(1 - \frac{C_w}{C_d}\right) \tag{7.13}$$

Using Eqn (7.13) in Eqn (7.9) gives

$$C_{opt} = \frac{[100 - (100/0.7)(1 - C_w/C_d)][130 - 100(1 - C_w/C_d)]C_d}{2 \times 10^4}$$

$$+ \frac{C_w}{0.7}\left(1 - \frac{C_w}{C_d}\right) + C_f$$

$$= \frac{C_d}{2}\left[\frac{(C_w/C_d + 0.3)(C_w/C_d - 0.3) + 2(C_w/C_d)(1 - C_w/C_d)}{0.7}\right.$$

$$\left. + \frac{2C_f}{C_d}\right]$$

The cost of daily energy (C_o), without using the wind turbine, is obtained by setting $x = 0$ in Eqn (7.9):

$$C_o = 1.3\frac{C_d}{2} + C_f$$

Therefore, the energy cost will be reduced only if

$$C_o - C_{\text{opt}} > 0 \tag{7.14}$$

or

$$1.3 - \frac{(C_w/C_d + 0.3)(C_w/C_d - 0.3) + 2(C_w/C_d)(1 - C_w/C_d)}{0.7} > 0 \tag{7.15}$$

which, on simplification, reduces to

$$\left(1 - \frac{C_w}{C_d}\right)^2 > 0$$

or

$$\frac{C_w}{C_d} < 1$$

Summary

Local power systems composed of solar, wind, and battery sets can provide year-round reliable supply to off-grid loads. Depending on the application, diesel engines can be considered as backup. Various configurations connecting the sources are possible, and these are site- and requirement-specific.

This chapter discusses schemes intended for combined operation of wind, solar, diesel, and battery systems—known as *hybrid systems*. The chapter first draws attention to the necessity of running hybrid systems for off-grid loads for economical reasons. Then a brief study of diesel engine characteristics and photovoltaic power generation, including maximum power point tracking, is presented. Various hybrid power system configurations—(a) wind–diesel, (b) wind–diesel with battery backup, and (c) wind–photovoltaic with battery backup—are presented and discussed, along with their merits and limitations. Any hybrid system should duly take into account the extent of availability of energy sources, the capital cost, and the operating and maintenance costs of the system components with a view to minimizing the production cost of electricity, while meeting the requirements of load and reliability of supply.

Problem

1. A three-phase permanent magnet (PM) generator has an open-circuit, line-to-neutral voltage of 50 V at 150 rpm and an inductance of 18 mH. The PM generator is driven directly by a wind turbine and feeds a diode-bridge rectifier. The rectifier output is connected to a 110-V battery bank and a 1.5-kW resistive load. In the absence of wind power, the battery can sustain the load for 2 hrs. The power equation for the wind turbine of radius 3 m is $P_W = 1.96R^2v^3$ and its C_p-λ data are the following.

λ	C_p
2.0	0.045
2.5	0.080
3.0	0.165
3.5	0.250
4.0	0.30
4.5	0.330
5.0	0.350
5.5	0.365
6.0	0.36
6.5	0.335
7.0	0.31
8.0	0.205
9.0	0.085
9.5	0.0

Find the critical wind speed below which the battery will start supplying the load.

Bibliography

Al Jabri, A.K. and A.I. Alolah 1990, 'Capacitance requirement for isolated self-excited induction generator', *IEE Proceedings*, vol. 137, pt B, pp. 154–9.

Al Jabri, A.K. and A.I. Alolah 1990, 'Limits on the performance of the three-phase self-excited induction generators', *IEEE Transactions on Energy Conversion*, vol. 5, pp. 350–6.

Arul Daniel, S. and N. Ammasai Gounden 2004, 'A novel hybrid isolated generating system based on PV fed inverter-assisted wind-driven induction generators', *IEEE Transactions on Energy Conversion*, vol. 19, no. 2, pp. 416–22.

Baliga, B.J. 1998, Evolution of MOS-bipolar power semiconductor technology', *Proceedings of the IEEE*, April 1988, pp. 409–18.

Bhadra, S.N. et al. 1996, 'Study of voltage build up in a self-excited, variable speed induction generator/static inverter system with dc side capacitor', *International Conference on Power Electronics Drive and Energy Systems for Industrial Growth (PEDES '96)*, New Delhi, pp. 964–70.

Bialasiewicz, J.T., E. Muljadi, S. Drouilhet, and G. Nix 1998, 'Hybrid power systems with diesel and wind turbine generation', *Proceedings of the American Control Conference, Philadelphia*, pp. 1705–09.

Bleijs, J.A.M., A.W.K. Chung, and A.J. Ruddell 1995, 'Power smoothing and performance improvement of wind turbine with variable speed operation', *Proceedings of the 17th BWEA Conference, London*, 1995, pp. 353–8.

Borowy, B.S. and Z.M. Salameh 1994, 'Optimum photovoltaic array size for a hybrid wind/PV system', *IEEE Transactions on Energy Conversion*, vol. 9, pp. 482–87.

Cadirci, I. and M. Ermis 1992, 'Double output induction generator operating at subsynchronous and supersynchronous speeds: steady state performance optimisation and wind-energy recovery', *IEE Proceedings*, vol. 139, pt B, pp. 429–42.

Chakraborty, C., S.N. Bhadra, and A.K. Chattopadhyay 1998, 'Excitation requirements for stand alone three-phase induction generator', *IEEE Transactions on Energy Conversion*, vol. 13, pp. 358–65.

Chan, T.F. 1992, 'Capacitance requirements of self-excited induction generators', *IEEE Transactions on Energy Conversion*, vol. 8, pp. 304–11.

Chen, Z. and E. Spooner 1998, 'Grid interface options for variable-Speed permanent magnet generators', *IEE Proceedings on Elctrical Power Applications*, vol. 145, pp. 273–83.

Chen, Z. and E. Spooner 1998a, 'Wind turbine power converters: A comparative study', *IEE International Conference, PEVD '98, London*, pp. 471–6.

Chen, Z. and E. Spooner 2001, 'Grid power quality with variable speed wind turbines', *IEEE Transactions on Energy Conversion*, vol. 16, no. 2, pp. 148–54.

Deltmer, R. 1990, 'Revolutionary energy—A wind/diesel generator with flywheel storage', *IEE Review*, April, pp. 149–51.

Demoulias, C.S. and P. Dokopolous 1996, 'Electrical transients of wind turbines in a small power grid', *IEEE Transactions on Energy Conversion*, vol. 11, pp. 636–42.

Dixon, J.W., and B. T. Ooi 1998, 'Indirect current control of a unity power-factor sinusoidal current boost type three-phase rectifier', *IEEE Transactions on Industrial Electronics*, vol. 35, pp. 508–15.

Enslin, J.H.R., M.S. Wolf, D.B. Snyman, and W. Swiegers 1997, 'Integrated photovoltaic maximum power point tracking converter', *IEEE Transactions on Industrial Electronics*, vol. 44, pp. 769–73.

Fitzgerald, A.E. and C. Kingsley, Jr 1961, *Electric Machinery—The Dynamics and Statics of Electromechanical Energy Conversion*, 2nd edn, McGraw-Hill Book Company, New York.

Freris, L.L. 1990, *Wind Energy Conversion Systems*, Prentice-Hall, UK.

Golding, E.W. 1955, *The Generation of Electricity by Wind Power*, E. & FN Spon Ltd, London.

Gyugyi, L. 1979, 'Reactive power generation and control by thyristor circuits', *IEEE Transactions on Industrial Applications*, vol. IA-15, no. 5, pp. 521–32.

Gyugyi, L. 1988, 'Power electronics in electric utilities: Static Var compensators', *Proceedings of the IEEE*, vol. 76, pp. 483–98.

Honorati, O., G.L. Bianco, F. Mezzetti, and L. Solero 1996, 'Power electronic interface for combined wind/PV isolated generating systems', *European Union Wind Energy Conference, Goteberg*, pp. 321–4.

Infield, D.G., G.W. Slack, N.H. Lipman, and P.J. Musgrove 1983, 'Review of wind/diesel strategies', *IEE Proceedings*, vol. 130, pt A, pp. 613–19.

Jadric, I., D. Borjevic, and M. Jadric 1997, 'A simplified model of a variable speed synchronous generator loaded with diode rectifier', *Power Electronics Specialist Conference*, pp. 497–502.

Jaydev, J. et al. 1995, 'Harnessing the wind', *IEEE Spectrum*, November, pp. 78–83.

Jenkins, N. 1993, 'Engineering wind farms', *Power Engineering Journal*, vol. 7, no. 2, pp. 53–60.

Jenkins, N. 1995, 'Some aspects of the electrical integration of wind turbines', *Proceedings of the BWEA Conference, London*, pp. 149–54.

Johnson, G.L. 1985, *Wind Energy Systems*, Prentice-Hall, Inc., Englewood Cliffs, NJ.

Jones, R. 1997, 'Power electronic converters for variable speed wind turbines', The Institution of Engineers report no. RP 1/8-8/8.

Koutroulis, E., K. Kalaitzakis, and N.C. Voulgaris 2001, 'Development of micro-controller-based photovoltaic maximum power point tracking control system', *IEEE Transactions on Power Electronics*, vol. 16, pp. 46–54.

Krause, P.C. and O. Wasynczuk 1998, *Electromechanical Motion Devices*, McGraw-Hill Book Company, New York.

Lander, C.W. 1993, *Power Electronics*, 3rd edn, McGraw-Hill Book Company, London.

Linders, J. 1989, 'Evaluation of an autonomous wind–diesel system with dynamic battery storage', *EWEC '89, Glasgow*.

Mani, Anna 1990, *Wind Energy Resource Survey of India*, Allied Publishers Ltd, New Delhi, pp. 347–52.

Manwell, J.F., J.G. McGowan, and B.H. Baily 1991, 'Electrical/mechanical options for variable speed wind turbines', *Solar Energy*, vol. 46, no. 1, pp. 41–51.

Mohan, N., T.M. Undeland, and W.P. Robbins 1995, *Power Electronics: Converters, Applications and Design*, 2nd edn, John Wiley & Sons, New York.

Murthy, S.S., O.P. Malik, and A.K. Tandon 1982, 'Analysis of self-excited induction generators', *IEE Proceedings*, vol. 129, pt C, pp. 260–65.

Musgrove, P.J. 1983, 'Wind energy conversion—An introduction', *IEE Proceedings*, vol. 130, pt A, pp. 506–16.

Nayar, C.V., W.B. Lawrence, and S.J. Phillips 1989, 'Solar/wind/diesel, hybrid energy systems for remote areas', *Proceedings of the 24th Intersociety Energy Conversion Engineering Conference, IECEC '89, Washington,* pp. 2029–34.

Pena, R., J.C. Clare, and G.M. Asher 1996, 'Doubly fed induction generator using back-to-back PWM converters and its application to variable speed wind energy generation', *IEE Proceedings Electric Power Applications,* vol. 143, no. 3, pp. 231–41.

Prasad, A.R., P.D. Ziogas, and S. Manias 1989, 'An active power-factor correction technique for three-phase diode rectifier', *IEEE Power Electronics Specialist Conference (PESC) Record,* pp. 58–66.

Raina, G. and O.P. Malik 1985, 'Variable speed wind energy conversion using synchronous machine', *IEEE Transactions on Aerospace and Electronic Systems,* vol. AES-21, pp. 100–4.

Rajakaruna, S. and R. Bonert 1993, 'A technique for the steady-state analysis of a self-excited induction generator with variable speed', *IEEE Transactions on Energy Conversion,* vol. 8, pp. 757–61.

Ramakumar, R. 'Wind energy systems', *IEEE Power Engineering Review,* pp. 6–9.

Ramakumar, R. and J.E. Bigger 1993, 'Photovoltaic systems', *Proceedings of the IEEE,* vol. 81, no. 3, pp. 365–77.

Richardson, R. and G.M. McNerney 1993, 'Wind energy systems', *Proceedings of the IEEE,* vol. 81, no. 3, pp. 378–89.

Saad-Saoud, Z., L.M. Craig, and N. Jenkins 1995, 'Static Var compensators for wind energy applications', *Proceedings of BWEA Conference, London,* pp. 347–52.

Salameh, Z.M. and L.F. Kazda 1986, 'Analysis of the steady state performance of the double output induction generator', *IEEE Transactions on Energy Conversion,* vol. EC-1, pp. 26–32.

Schwartz, R.J. 1993, 'Photovoltaic power generation', *Proceedings of the IEEE,* vol. 81, no. 3, pp. 355–64.

Scott, G.W., W.F. Wilreker, and R.K. Shattons 1984, 'Wind turbine generator interactions with diesel generators on an isolated power system', *IEEE Transactions on Power Apparatus and Systems*, vol. PAS-103, pp. 933–7.

Silva, S.R. and R.O.C. Lyra 1993, 'PWM converter for excitation of induction generators', *Proceedings of the Fifth European Conference on Power Electronics and Applications (EPE '93), Brighton, UK*, pp. 174–8.

Slemon, G.R. 1987, 'Modelling of induction machines for electric drives', *IEEE Transactions on Industrial Applications*, vol. IA-25, pp. 1126–31.

Sloot, I.G. and E. de Vries 2003, 'Inside wind turbines— Fixed vs variable speed', *Renewable Energy World*, pp. 30–40.

Smith, G.A. 1993, 'High quality mains power from variable-speed wind turbines', *Renewable Energy, Conference Publication Number 385*.

Smith, G.A. et al. 1993, 'A sine wave interface for variable speed wind turbines', *Conference Record*, the European Power Electronics Association, pp. 97–102.

Smith, G.A. and K.A. Nigim 1981, 'Wind-energy recovery by a static Scherbius induction generator', *IEE Proceedings*, vol. 128, pt C, pp. 317–24.

Spooner, E., and A.C. Williamson 1992, 'Permanent Magnet Generators for Wind Power Applications', *Proceeding of the 1992 International Conference on Electrical Machines, Manchester*, pp. 1048–52.

Spooner, E., and A.C. Williamson 1996, 'Direct coupled permanent magnet generators for wind turbine applications', *IEE Proceedings on Electrical Power Applications*, vol. 143, no. 1, pp. 1–8.

Spooner, E. and B.J. Chalmers 1992, 'Torus: A slotless, toroidal-stator, permanent-magnet generator', *IEE Proceedings on Electric Power Applications*, vol. 139, no. 6, pp. 497–506.

Stavrakakis, G.S. and G.N. Kariniotakis 1995, 'A general simulation algorithm for the accurate assessment of isolated diesel–wind turbine systems interaction, Pt I: A general multi-machine power system Model', *IEEE Transactions on Energy Conversion*, vol. 10, pp. 577–83.

Tande, J.O.G. 1998, 'Impact of wind turbines on voltage quality', *Proceedings of the 8th International Conference on Harmonics and Quality of Power, ICHQP '98, Athens*, pp. 1158–61.

Tang, Y. and L. Xu 1995, 'A flexible active and reactive power control strategy for a variable speed constant frequency generating system', *IEEE Transactions on Power Electronics*, vol. 10, pp. 472–8.

Tsitsovits, A.J. and L.L. Freris 1983, 'Dynamics of an isolated power system supplied from diesel and wind', *IEE Proceedings*, vol. 130, pt A, pp. 587–95.

Verdello, P. and G.D. Marques 1993, 'Decoupled model of PWM voltage converter connected to the AC mains', *IECON '93*, pp. 2/1021–6.

Warne, D.F. and P.G. Kalnan 1997, 'Generation of electricity from wind', *Proceedings of the IEE*, vol. 124, no. 11R, pp. 963–85.

Watnobe, C.H. and A.N. Barreto 1987, 'Self-excited induction generator/force commutated rectifier system operating as a dc power supply', *IEE Proceedings*, vol. 134, pt B.

Watson, D.B. 1998, 'Circuit model and self-excitation for the induction generator', *International Journal of Electrical Engineering Education*, vol. 25, Manchester, pp. 163–70.

Wilmshurst, S.M.B. 1988, 'Control strategies for wind turbines', *Wind Engineering*, vol. 12, no. 4, pp. 236–49.

Wortmann, F.X. 1978, 'Airfoil profiles for wind turbines,' Institüt für Aerodynamik und Gasdynamik, Universität Stuttgart, Report no. 78–9.

Xu, X., R. De Doncker, and D.W. Novotny 1988, 'A stator flux oriented induction machine drive', *PESC '88 Record*, pp. 870–6.

Index

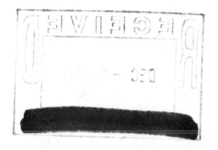